About the Author

ROBERT ADAIR is Sterling Professor Emeritus of Physics at Yale University and a member of the National Academy of Sciences. His research has largely been concerned with the properties of the elementary particles and forces of the universe.

THE PHYSICS
OF BASEBALL

Third Edition, Revised, Updated, and Expanded

ROBERT KEMP ADAIR

Perennial

An Imprint of HarperCollinsPublishers

THE PHYSICS OF BASEBALL (Third Edition, Revised, Updated, and Expanded). Copyright © 1990, 1994, 2002 by Robert K. Adair. All rights reserved. Printed in the United States of America. No part of this book may be used or reproduced in any manner whatsoever without written permission except in the case of brief quotations embodied in critical articles and reviews. For information address Harper-Collins Publishers Inc., 10 East 53rd Street, New York, NY 10022.

HarperCollins books may be purchased for educational, business, or sales promotional use. For information please write: Special Markets Department, Harper-Collins Publishers Inc., 10 East 53rd Street, New York, NY 10022.

Library of Congress Cataloging-in-Publication Data

Adair, Robert Kemp.
    The physics of baseball/ Robert Kemp Adair.—3rd ed.
        p. cm.
    Includes index.
    ISBN 0-06-008436-7
    1. Physics. 2. Baseball. 3. Force and energy. I. Title.

QC26 .A23   2002
796.357'01'53—dc21

                                        2001039886

02 03 04 05 06 WB/RRD 10 9 8 7 6 5 4 3 2 1

To the memory of my grandfather Theodore Wiegman (1876–1953), who often sat with a small boy behind third base watching 3-I League baseball in the old Fort Wayne ballpark in the 1930s, and to the memory of my son, James Cleland Adair (1957–1978), graceful first baseman on his championship little league team.

CONTENTS

LIST OF FIGURES

PREFACE

Late in the summer of 1987, Bart Giamatti, then President of the National Baseball League and later Commissioner of Baseball, an old friend and colleague of mine from his days as professor of English and then president of Yale, asked me to advise him on the elements of baseball that might be best addressed by a physicist. I told Bart that I would be delighted to do so—and that I expected that I would have so much fun at such a job that I would find it wrong to accept any payment for it. Bart, ever the English professor and sensitive to words, responded by appointing me "Physicist to the National League"—a title that absolutely charmed the ten-year-old boy that I hope will always be a part of me.

As I considered the few questions Bart posed to me, it became increasingly clear that to answer them properly, I had better understand almost all baseball as best I could. Hence, more as a delightful hobby than as an obligation, I attempted to describe quantitatively as much of the action of baseball as I could in a report to Bart Giamatti as League President. He then suggested that I publish it, suitably expanded, as a book. I was pleased that Bart saw the final manuscript before his death, and that he liked it.

Written for fun, and originally for Bart Giamatti, this book is not meant as a scholarly compendium of research on baseball—though I have borrowed extensively from the work of others. Moreover, it is not meant as the definitive treatise on the aspects of baseball that it considers. I have not hesitated to make best estimates on matters that I do not understand as well as I would like, and while I believe that I have made no egregious blunders, the physics of baseball is far from trivial and I surely will have slipped somewhere.

Although I found no errors in the first and second editions that were central to baseball, over the past decade I have learned more about the sport. Hence, I have made a number of modest changes in numbers and text in this third edition to reflect my improved understanding of the game, and have added comments on points that I had not previously addressed. These changes modify, but do not radically change, the analyses presented in the first and second editions.

Aside from such corrections and additions, in this third edition I have also added some new material on matters that I found of interest to myself, fans, and players. In particular I have material borrowed from neurophysiology that I believe illuminates the pitcher-batter duel central to the game. I have also incorporated some of the elegant researches by others that have better defined the collision of the ball with wooden bat, and have discussed home runs and the "lively ball" at greater length.

I have written this book for those interested in baseball, not in the principles of physics, and therefore I have slighted descriptions of the details of the calculations to which I refer. These calculations were usually conducted by simple BASIC programs on a personal computer. The few formulas which are included, segregated as Technical Notes at the end of each chapter, are introduced to provide, for those who might be especially interested, a succint description of the models I used.

Robert Kemp Adair
Hamden, Connecticut

THE PHYSICS OF BASEBALL

MODELS AND THEIR LIMITATIONS

A small, but interesting, portion of baseball can be understood on the basis of physical principles. The flight of balls, the liveliness of balls, the structure of bats, and the character of the collision of balls and bats are a natural province of physics and physicists.

In his analysis of a real system, a physicist constructs a well-defined model of the system and addresses the model. The system we address here is baseball. In view of the successes of physical analyses in understanding arcane features of nature—such as the properties of the elementary particles and fundamental forces that define our universe (my own field of research) and the character of that universe in the first few minutes of creation—it may seem curious that the physics of baseball is not at all under control.

We cannot calculate from first principles the character of the collision of an ash bat with a sphere made up of layers of different tightly wound yarns, nor do we have any precise understanding of the effect of the airstream on the flight of that sphere, with its curious yin-yang pattern of stitches. What we can do is construct plausible models of those interactions that play a part in baseball that do not violate basic principles of mechanics. Though these basic principles—such as the laws of the conservation of energy and momentum—severely constrain such models, they do not completely define them. It is necessary that the models touch the results of observations—or the results of the controlled observations called experiments—at some points so that the model can be more precisely defined and used to interpolate between known results, or to extrapolate from them.

Baseball, albeit rich in anecdote, has not been subject to extensive quantitative studies of its mechanics—hence, models of baseball are not as well founded as they might be.* However connected with experience, model and system—map and territory—are not the same. The physicist can usually reach precise conclusions concerning the character of the model. If the model is well chosen, so as to represent the salient points of the real system adequately, conclusions derived from an analysis of the model can apply to the system to a useful degree. Conversely, conclusions—although drawn in a logically impeccable manner from premises defined precisely by the model—may not apply to the system if the model is a poor map of the system.

Hence, in order to consider the physics of baseball, I had to construct an ideal baseball game which I could analyze that would be sufficiently close to the real game so that the results of the analysis would be useful. The analysis was easy, the modeling was not. I found that neither my experience playing baseball (poorly) as a youth nor my observations of play by those better fitted for the game prepared me for the task of constructing an adequate model of the game. However, with the aid of seminal work by physicists Paul Kirkpatrick, Lyman Briggs, and others,† and

---

*More effort has been spent on analyses of golf, probably because there are economic advantages to the support of research by manufacturers, who might make and sell better balls and clubs. Baseballs and wooden bats used in professional baseball are made to specifications set down by major league officials and are less subject to manufacturers' improvements.

†The papers by Briggs and Kirkpatrick were published in the *American Journal of Physics*, Briggs's paper in 1959 and Kirkpatrick's in 1963. There is a small literature on the physics of baseball. As with most fields of science, much of this work is incorrect. Science is difficult, and original published results—the raw materials of science, as yet untouched by the sifting and winnowing process that results in reliable knowledge—are notoriously unreliable. In this book, I try to report "sifted" results and to give the reader some idea of uncertainties as I see them.

with help from discussions with other students of the game, such as my longtime associate R. C. Larsen, I believe that I have been able to arrive at a sufficient understanding of baseball so that some interesting conclusions, drawn from analyses of my construction of the game, are relevant to real baseball.

In all sports analyses, it is important for a scientist to avoid hubris and to pay careful attention to the athletes. Major league players are usually serious people, intelligent and knowledgeable about their craft. Specific, operational conclusions held by a consensus of players are seldom wrong. However, since baseball players are athletes, not engineers or physicists, their analyses and rationale may be imperfect. If players think that they hit better after illegally drilling a hole in their bat and filling it with cork, they must be taken seriously, though the reasons they give for their "improvement" may not be valid.

I hope that nothing in the following material will be seen by a competent player of the game to be definitely contrary to his experience in playing the game. Honed by a century of intelligent trial and error, baseball must surely be played correctly — though not everything *said* about that play, by players and others, is impeccable. Hence, if a contradiction arises concerning some aspect of my analyses and the way the game is actually played, I would presume it likely either that I have misunderstood that aspect myself or that my description of my conclusion was inadequate and subject to misunderstanding.

Just as the results discussed here follow from analyses of models that can only approximate reality, the various conclusions have different degrees of reliability. Some results are quite reliable: The cork, rubber, or whatever, stuffed into holes drilled in bats certainly does not increase the distance the bat hits the ball. Some results are hardly better than carefully considered guesses: How much does backspin affect the distance a long fly ball travels? Although I have tried to convey the degree of reliability of different conclusions, it may be difficult to evaluate the caveats properly. By and large, the qualitative results are usually reliable, but most of the quantitative results should be considered with some reserve, perhaps as best estimates.

In spite of their uncertainties, judiciously considered quantitative estimates are interesting and important. Whatever their uncertainties, they often supplant much weaker, and sometimes erroneous, qualitative insights. Consequently, I have attempted to provide numerical values almost everywhere: sometimes when the results are somewhat uncertain, sometimes when the numbers are quite trivial but not necessarily immediately accessible to the reader.

As this exposition is directed toward those interested in baseball, not physics, I have chosen to present quantitative matters in terms of familiar units. Hence, I use the English system of measures—distances in feet and inches, velocities in miles per hour (mph), and forces in terms of ounce and pound weights. Moreover, I have often chosen to express effects on the velocities of batted balls in terms of deviations of the length of a ball batted 400 feet (likely to be a long home run) under standard conditions.

To express the goals of this book, I can do no better than to adopt a modification of a statement from Paul Kirkpatrick's article "Batting the Ball": *The aim of this study is not to reform baseball but to understand it.* As a corollary to this statement of purpose, I must emphasize that the book is not meant as a guide to players; for of all of the ways to learn to better throw and bat a ball, an academic study of the mechanics of the actions must be the least useful.

2

THE FLIGHT OF
THE BASEBALL

THE BASEBALL—AIR RESISTANCE

From the *Official Baseball Rules: 2001*:

1.09 The ball should be a sphere formed by yarn wound around a small sphere of cork, rubber, or similar material covered with two stripes of white horsehide or cowhide, tightly stitched together. It shall weigh not less than 5 nor more than 5¼ ounces avoirdupois and measure no less than 9 nor more than 9¼ inches in circumference.

The description of the baseball in the rule book, ingenuous and charming, is not that of an engineer; the manufacturer (once in Chicopee, Massachusetts, then Haiti, then Taiwan, and now, at the beginning of the third millennium, in Costa Rica) is given these further directions: "The cork-rubber composite nucleus, enclosed in rubber, is wound with 121 yards of blue-gray wool yarn, 45 yards of white wool yarn, and 150 yards of fine cotton yarn. Core and winding are enclosed by rubber cement and a two piece cowhide—horsehide before 1974—cover hand-stitched together with just 216 raised red cotton stitches."

Much more is required to completely define the ball that is the center of the sport of baseball, but its flight is largely determined by the size and weight constraints listed in the rules. The paths of baseballs projected at velocities common to the game are strongly influenced by air resistance. As the ball passes through the air, it pushes the air aside and loses energy, and thus velocity, through the work it does on the air. The forces on the ball from the resistance of the air are typically of the same magnitude as the force of gravity. A ball batted with an initial velocity of 110 mph at an angle of 35° from the horizontal would go about 750 feet in a vacuum; at Shea Stadium in New York, it will travel only about 400 feet. Hence, it is necessary to understand the fluid dynamics of air flow around spheres to understand the flight of a baseball.

When an object (such as a baseball) passes through a fluid (such as air), the fluid affects the motion of the object as it flows about that object. Moreover, for all fluids and all objects, the character of the flow of the fluid is largely determined by the value of a (dimensionless) Reynolds number proportional to the density of the fluid, the fluid velocity, and the size of the object, and inversely proportional to the viscosity of the fluid.[a] For a given Reynolds number, the behavior of the gaseous fluid of stars— interacting with each other through gravity—that make up a galaxy a hundred thousand light-years across is described in very much the same way as the behavior of the molecules of air passing through an orifice 1 micron across, where a micron is about equal to the resolution of a high-power microscope.

The most interesting actions in the game of baseball take place when velocities of the ball range from a few miles per hour (and Reynolds numbers of 10,000) to values near 120 mph (and Reynolds numbers near 200,000). For velocities in that range below about 50 mph, the flow of the air around the ball is rather smooth, though trailing (Von Karman) vortices are generated. This airflow does not actually reach the surface of the ball where there is a quiet (Prandtl) boundary layer. A very, very small insect (perhaps a plant aphid) sitting on the moving ball would feel no breeze at all. At velocities above 200 mph the flow penetrates the boundary layer (the aphid would have to hold on very tightly to avoid being blown off) and the air at the boundary—and trailing behind the ball—is quite turbulent. I label the two regions conveniently (if a little inaccurately) as *smooth* and *turbulent*.

Hence, for a baseball passing through air at a velocity less than 50 mph the airflow is smooth, while the airflow is turbulent for velocities greater than 200 mph. But much of the subtlety of baseball is derived from the fact that so much of the game is played in the region between definitely smooth flow and definitely turbulent flow, at ball velocities greater than 50 mph and less than 120 mph. For balls traveling at the transition velocities between 50 and 120 mph, the flow can be smooth or turbulent, depending on the detailed character of the surface of the ball and its motion. By and large, turbulence will be induced at lower velocities by roughness in the surface, and held off to higher velocities if the surface is very smooth. Surprisingly, at a given velocity the air resistance is *less* for turbulent flow than for smooth flow. It seems that at low velocities the ball, with its boundary layer of still air, is effectively larger than it is at higher velocities with the boundary layer blown off and thus the higher-velocity ball moves a smaller column of air.

From our understanding of fluid flow, it is convenient to describe the drag or retarding force on a moving baseball[b] (or equivalent sphere) as proportional to the cross-sectional area of the ball, as a larger ball must push more air out of the way. The force is also proportional to the square of the velocity of the ball (doubling the velocity increases the drag by a factor of four). A ball with double the velocity must push twice as much air out of its way, and that air will be pushed twice as hard. If the air is less dense, it is easier to push away; hence the drag is also proportional to the density of the air. Consequently, the drag varies to some extent with temperature and altitude, just as the air density varies with those factors. As I have mentioned, the character of the airflow around the ball can change with velocity, and such changes affect the resisting drag force also. We take that into account by a further proportionality of the drag to a "drag coefficient" that depends only on the value of the Reynolds number which is proportional to the velocity of the ball.

Figure 2.1 shows an estimate of the variation of the drag coefficient for a baseball as a function of the velocity of the ball. The drag force on the baseball will also depend to some extent upon the orientation of the stitches on the ball. When the ball is rotating — as is usually the case — the drag will depend on the position of the axis of rotation with respect to the stitch pattern of the ball, on the direction of the axis with respect to the ground and the direction of the ball's flight, and on the velocity of rotation of the ball. The drag on a rapidly spinning ball is probably slightly larger than that on a slowly rotating ball. But that effect must be small; I estimate that the incremental drag on the spinning ball will usually not be much larger than 5 percent of the drag on the nonspinning ball. This would mean that

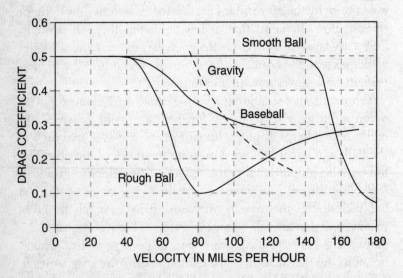

FIGURE 2.1: *The drag coefficient vs. ball velocity for various balls the size of a baseball at sea level and a temperature of 68° F. The broken curve suggests the values of the coefficient required to produce a drag force equal to gravity.*

for a 90-mph fastball thrown with a spin of 1500 rpm, the extra drag will reduce the speed of the ball so that it will cross the plate traveling about 0.5 mph slower than a ball thrown with very little spin.

For rotating balls, the dependence of the drag upon these factors is not likely to be large, however, and we can consider that the drag effects described here represent a kind of average over different orientations of the stitch patterns with the spin directions. Aside from these caveats, the values of the coefficients for a baseball are not well known, but wind tunnel measurements have been made of the forces on balls at velocities up to 95 mph that support the solid curve. In particular, wind tunnel measurements showed the ball suspended nearly motionless in an upward-directed 95-mph airstream. Therefore, for a ball moving through the air with a velocity of 95 mph, the drag force is about equal to the weight of the ball. Hence, the terminal velocity of a ball dropped from a great height is but 95 mph.

The broken-curve line in the figure corresponds to values of the drag coefficient that would generate a drag force equal to the force of gravity. The ball suspended in the 95-mph vertical airstream was held steady by the force of the air. This would be the case only if the force is greater if the ball is falling, and hence moving faster through the air, and if the force is weaker if the ball is rising, and thus moving more slowly with respect to the air. Consequently, the observed stability of the ball demands that the solid curve cross the broken line from left to right at 95 mph, further defining the variation of the drag coefficient with the velocity of the baseball.

The values expressed for larger velocities, of up to 150 mph, are estimates, albeit guided by theoretical considerations. The mean uncertainties are perhaps 10 percent for velocities less than 120 mph, which is near the highest velocity reached by a ball in play. The rather gradual reduction in the drag coefficient with increasing velocity, from the value of about 0.5 for velocities less than 50 mph to those of about 0.2 for velocities greater than 120 mph, suggests that the transition from smooth to turbulent flow of the air passing the baseball in flight occurs gradually.

The values of the drag coefficient for an ideally smooth ball and a uniformly rough ball—about as rough as a ball completely covered with stitches—are shown also. The variation of the drag coefficient with velocity will have the same general character for a ball that is uniformly a little rougher or a little smoother, but the drag minimum will be found at lower velocities for a rougher ball and at higher velocities for the smoother one.*

At the velocities of 50 to 120 mph dominant in baseball, the air passes over a smooth ball the size of a baseball in a smooth high-resistance flow; turbulence is not induced until velocities approach 200 mph. However, a real baseball, roughened by its raised-stitch patterns, induces low-resistance turbulent flow at baseball velocities. Consequently, if the baseball were quite smooth rather than laced with protuberant stitches, as is the case, it could not be thrown or batted nearly as far as it in fact is—a stitched baseball batted 400 feet would travel only about 350 feet, if the ball were very smooth. This effect is dramatic in golf; the air resistance from the smooth flow about a smooth ball is so great that the ball goes nowhere. The ball is therefore artificially roughened by the dimples impressed in the covering to induce turbulence and reduce air resistance.

At velocities near 175 mph, where the resistive force for smooth balls falls off sharply, the resistance on the ball actually becomes smaller as the velocity of the ball increases. The resistance on a smooth ball the size of a baseball traveling 190 mph is less than the resistance on such a ball traveling 160 mph. This sharp dip in the drag coefficient at the onset of turbulence has been called the "drag crisis." For uniformly rough balls the sharp reduction in drag with increasing velocity comes at lower velocities; the rougher the ball the lower the velocity at which turbulence is induced and the lower velocity

---

*These properties of balls of various degrees of roughness were established experimentally by E. Aschenbach and published in 1974 in the *Journal of Fluid Mechanics.*

of the drag crisis. If a baseball suffered such a drag crisis at velocities typical of the game, the anomalous effects could be important.* For example, a ball hit against the wind might go farther than a ball hit with the wind. It seems, however, that the baseball, usually rotating in flight and thus presenting different configurations of smoothness and roughness to the air as a consequence of the changing orientation of its stitches, is by no means a *uniformly* rough ball and therefore does not undergo the sudden transition from smooth to turbulent flow that characterizes a drag crisis.

Since the drag for a rough ball can be less than that for a smoother ball, even if there is no drag crisis for a baseball the distance a batted ball will travel might still depend upon the character of its surface. A very rough, scarred ball with a surface that could induce turbulence at low velocities could well travel farther than a new, smooth ball. Smaller changes could also be significant. Such a change in the surface of the ball may have occurred in 1974, when the traditional horsehide cover was replaced by cowhide. However, a judicious estimate suggests that for the range of surface conditions tolerated for baseballs used in the major leagues, the dependence of the drag on the character of the basic skin surface can be neglected; the skins of cows and horses are not that different. Still, any significant changes in the height of the stitches might change the velocity at which turbulence begins and thus affect the flight of the ball and the distance it can be hit.

Returning to the regulation baseball, the solid curve of Figure 2.2 shows the variation of the drag force on the ball with velocity derived from the baseball drag coefficients of Figure 2.1. The force is expressed in terms of the weight of the ball; hence, the value 1.0 of the ordinate corresponds to the force of gravity. Again, there may be significant differences in the drag force for different orientations of the stitch configurations with respect to

---

*Cliff Frohlich discusses the drag on baseballs — and much else about baseball — in an article in the *American Journal of Physics* published in 1984.

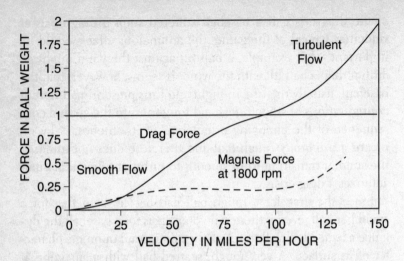

FIGURE 2.2: *The solid line shows the variation with velocity of the drag force on a baseball. The broken line shows the variation with velocity of the Magnus force for a ball spinning at a rate of 1800 rpm. The forces are expressed in units of the ball's weight.*

the direction of motion, hence, the forces shown here must be considered as a kind of average over the range of possible stitch orientations.

SPIN AND THE MAGNUS COEFFICIENT

Interest in the left-right curvature of balls sailing through the air is probably as old as ball games themselves. Isaac Newton,[*] at the age of twenty-three, discussed the curvature of court tennis balls—a little like small baseballs—in terms that make good sense today. In the nineteenth-century genesis of mathematical physics, Lord Rayleigh analyzed the curvature of the path of spinning balls and P. G. Tait, an eminent Scottish physicist, wrote

[*]Described in his article entitled "New Theory of Light and Colours," published in the first volume of the *Philosophical Transactions of the Royal Society of London* in 1672.

extensively on the curves of golf balls—perhaps in an attempt to understand and cure a slice.

The total force on a baseball, from the normal air pressure of 14.6 pounds per square inch, pushing the ball toward third base as it travels from pitcher to batter, is nearly 100 pounds. Of course, there is ordinarily a nearly identical force pushing the ball toward first base. If these forces differ by as much as an ounce and a half, or about one part in a thousand, a ball thrown to the plate at a velocity of 75 mph will be deflected, or curve, a little more than a foot. Such modest, asymmetric force imbalances are generated by a spinning ball and by asymmetric placement of the stitches on the ball.

If the resistive force on a ball is proportional to the square of the velocity of the air passing the ball, it would seem probable that there would be such an unbalanced force on a spinning ball, on an axis normal to the velocity even as the velocity through the air of one side of the ball at the spin equator is greater than the velocity of the other side. Such a force, directed at right angles to the direction of the air velocity and to the axis of spin, has long been known and is usually called the Magnus force.

Some insight into the force imbalance can be gained by considering the different forces that follow from the different velocities of the opposite sides of a spinning ball. Figure 2.3 suggests the origin of this transverse force for the "normal" situation, where the air resistance forces increase as the velocity increases. Consider that a curveball is thrown (from left to right in the figure) by a right-handed pitcher at a speed of 70 mph, so that it rotates 16 times counterclockwise (as seen from above) in its trip of about 56 feet from the pitcher's hand to the plate. Such a ball will be rotating at a rate of about 1800 rpm, i.e., about one-half the rate of a typical small synchronous electric motor. The side of the ball toward first base (at the top of the figure) then travels about 16 times 9 inches (the circumference of the ball), equal to 12 feet, less than 56 feet, while the side toward third base travels 12 feet farther than 56 feet. The velocity of the third-base side is then about 85 mph, while the velocity of the first-base side is only 55. As shown in Figure 2.2, the drag force increases with velocity; i.e., the difference in the pressure of the air on the front face of the moving ball is greater than on the rear face, and that pressure difference increases with velocity. We can then expect the drag force on the third-base side of the ball, which is traveling faster through the air, to be greater than the drag on the first-base side, which is traveling more slowly, and the ball will be deflected toward first base.

While a complete description of the Magnus force would surely be much more complex than the foregoing description, which follows Newton, for a simple and useful picture it seems to me to be relatively transparent and to contain as much of the truth as any other simplistic picture. Note that although the Magnus force probably does vary somewhat with the orientation of stitches with respect to the spin axis of the ball, the stitches do not contribute to the force in any simple way. Spinning smooth balls, such as those used in table tennis, curve with no aid from stitches.

If the resistive drag force varies only as the square of the velocity and the Magnus force is only an imbalance in that resistance which follows from the faster motion of one side of the ball through the air than the other, we should expect that the Magnus force would be proportional to spin frequency, proportional to

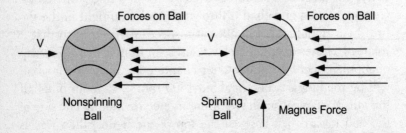

FIGURE 2.3: *The character of the drag forces on a nonspinning and spinning ball suggests the origin of the Magnus force. The imbalance in the components of forces normal to the surface of the spinning ball generates a force directed upward on the ball.*

the air velocity (or ball velocity),[*] and proportional to the value of the drag coefficient at that velocity.[c]

Though the Magnus force can be described qualitatively in terms of such an imbalance of the drag forces, a reliable quantitative description of the force on a stitched baseball is not available (or more properly, measurements that should be reliable do not agree). Hence, I adopt a model of the Magnus force on a baseball that I believe will reflect, correctly, the general characteristics of the real force as expressed by the drag imbalance — and that, at worst, is not likely to be seriously in error. So from this view the Magnus force, described by the imbalance of resistive forces on the ball that follows from the imbalance of velocity of the airflow past the spinning ball, is proportional to the rate of change of the drag resistance with velocity which is the slope of the drag resistance curve shown as the solid line in Figure 2.2. From this model[d] of the Magnus force, the magnitude of the force (expressed in units of the weight of the ball) is shown in Figure 2.2 as a broken line for balls rotating at a rate of 1800 rpm, which is near the maximum for thrown balls. I estimate that the uncertainties in my estimate of the magnitude of the Magnus force on the baseball, as plotted in Fig. 2.2, are about 25 percent.

Like the related drag force, the Magnus force must depend to some extent on the stitch configuration and thus the orientation of the spin axis. The values plotted in Figure 2.2, then, represent a kind of average over axes orientations.

According to this model, the maximum Magnus force on a ball spinning at a rate of 1800 rpm is seen to be about one-fourth of the weight of the ball. Hence we cannot expect a ball spinning at that rate to curve more than one-fourth the distance it will fall under gravity. Since the variation of velocity is proportional to the rate of spin, the Magnus force on a ball will be proportional to that spin rate; e.g., the force on a ball spinning at a rate of 900 rpm will be one-half that shown by the broken line in the figure.

---

[*]This relation was found in measurements by R. Watts and R. Ferrar and reported in the *American Journal of Physics* in 1987.

Note that the Magnus force on a ball spinning at a constant rate increases with velocity up to speeds near 60 mph—the speed of a curveball—and then probably levels off. According to this model of the force, for a given spin rate the Magnus force is no greater for a ball thrown at a velocity of 90 mph—such as a major league fastball—than it is for a 60-mph ball. We might expect, then, that a hard-thrown ball will curve much less than a slower pitch, though the spin rates might be the same. The Magnus force can be expected to be about the same for the fastball, but the force has less time in which to act, as the ball reaches the plate sooner.

The forces that cause the ball to be deflected must also generate a torque that slows down the spin.[e] For a hard-hit ball traveling with an initial velocity of 110 mph, I use the model of the Magnus force I have adopted to estimate that the spin rate will decrease at a rate of about 20 percent per second. For a 400-foot home run, the backspin applied by the bat (perhaps 2000 rpm) would be reduced to about 500 rpm when the ball lands about five seconds later. This is consistent with experience: Fly balls do not spin out of the gloves of outfielders when they barely catch the ball in the tip of their gloves—sometimes called "snow cone" catches—but balls do sometimes spin out of the gloves of catchers fielding foul pop-ups, whose terminal spin rates can be much greater.

DISTANCE OF FLIGHT OF A BATTED BALL

Whether it is projected by arm or bat, the distance a ball will travel can be calculated using the simple ballistic relations governing a body in flight and taking the retarding drag force in the direction of motion and the Magnus force, normal to the direction of motion, from the values shown in Figure 2.2. Typical trajectories are shown in Figure 2.4, and a graph of the maximum distance vs. the initial ball velocity is shown in Figure 2.5. From these calculations, the maximum distance is obtained with balls projected at an initial angle of about 35° from the horizontal, though balls projected at 30° or 40° travel almost as far. Note that the ball falls at a rather steep angle at the end of its flight; the trajectories are not symmetric. We find that the 385-foot fly ball hit at the optimum angle of about 35 degrees will be in the air for about five seconds; a really high fly ball—or pop fly—will stay in the air more than six seconds. To put this into perspective, an average right-handed batter will run from home to first in about 4.3 seconds.*

Obviously these trajectories and maximum distances depend upon the drag and the drag coefficients, which are imperfectly known. Assuming an uncertainty of 10 percent in the drag coefficient at high velocities, the uncertainty in the maximum distance of a ball's flight will be about 3 percent, or about 12 feet for a 400-foot home run. Since we are primarily interested in *differences* in distance that follow from changes in conditions, such an accuracy is acceptable.

As a ball is thrown or batted for distance in a real situation, it usually has appreciable backspin (on the order of 2000 rpm), which generates a significant force perpendicular to the direction of motion, generally upward. Though the spin will probably increase the drag on the ball slightly, as a consequence of the lift provided by backspin, the Magnus force will somewhat increase the distance traveled by balls hit or thrown at smaller angles and reduce the distance of balls hit or thrown at larger angles. In general, the backspin can be expected to produce a modest increase in the distance the ball will travel, as well as a decrease of a few degrees in the optimum angle of projection. According to my cal-

---

*A very fast left-handed batter will reach first base in about 3.7 seconds after a drag bunt, a few tenths of a second—a step—faster than an equally fast right-handed batter, while an unusually slow runner, such as catcher Ernie Lombardi of the 1940s, might take more than 5 seconds to go from home to first. In a timed racelike effort in 1921, Maurice Archdeacon circled the bases in 13.4 seconds and the legendary James "Cool Papa" Bell of the Negro Leagues is said to have made it in less than 13 seconds. To put these times into context, a world-class sprinter like Carl Lewis in the 1980s, running with track shoes on a carefully prepared track, would take about 11 seconds to run the 120 yards if the bases were laid out in a line. He would need about 11.6 seconds to run in a straight line the 127 yards that a reasonable course around the bases would measure.

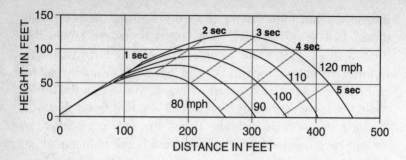

FIGURE 2.4: *The trajectories of balls projected at an angle of 35° with different velocities. The balls are assumed to be rotating with an initial backspin of about 1 revolution per 5 feet—or 1800 rpm for a ball traveling 100 mph. The solid circles show positions of the ball at intervals of one second.*

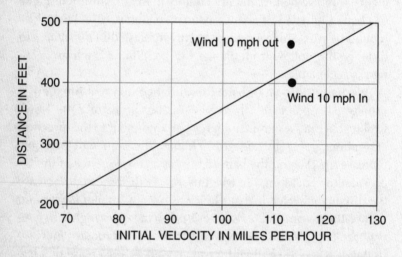

FIGURE 2.5: *The distance a baseball travels as a function of the initial velocity of the ball when projected at an angle of 35° with an initial backspin of 2000 rpm. The solid circles show the effects of a 10-mph wind blowing in and out.*

culations, a ball launched without backspin at an angle of 35° traveling 385 feet would travel roughly 400 feet if the backspin were about 2000 rpm. (Coming off of the bat, such a ball would rotate about once every five feet of flight.) There are two countering effects: The backspin causes the ball to stay in the air longer, thus increasing the distance it travels; but that spin also increases the drag, hence decreasing the flight distance. Since neither effect is known precisely, the magnitude of the backspin effect is uncertain.

Since the backspin itself depends in detail on the exact manner in which the ball is thrown or struck by the bat, to calculate the representative trajectories and distances given in Figures 2.4 and 2.5, I have assumed a backspin of about 1 rotation per 5 feet of flight. Thus the rotation rate is proportional to the velocity. For a ball traveling at 100 mph, that rate will be about 1800 rpm.

Obviously the distance a ball will travel is strongly affected by the wind. Figure 2.5 also shows the effect of winds of 10 mph (near the average velocity throughout the United States) on the flight of a 400-foot home run. With the wind behind the batter, the ball will go about 30 feet farther; with the wind against the batter, the 400-foot home run to center field will end up as 370-foot out.

Although the effects of soft breezes may seem small, the consequences are not necessarily trivial. My analyses of the distribution of fly ball distances and home runs suggest that for an average player, a 1 percent change in the distance the ball travels translates to about a 7 percent change in the probability of hitting a home run. I bring this estimate into our experience by considering that this ratio leads to about one ball being caught on the warning track, or landing in the first row of the stands, for every ten home runs. The ratio is not so large for the great home run hitters: Ruth, Maris, McGwire, and Sosa didn't hit many balls to the warning track. When they connected, the ball went out.

In general, ballparks are laid out so that the line from "home base through the pitcher's plate to second base shall run East-Northeast,"* in order that no batter would have to face a Nolan Ryan fastball in the late afternoon with the sun in his eyes. Hence, the prevailing westerlies of the Northern Hemisphere tend to blow out toward right-center field, helping the batter. While for most outdoor ball parks the effect of the wind down on

---

*Official Baseball Rules, Section 1.04.

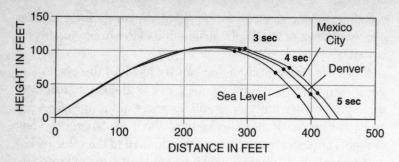

FIGURE 2.6: *Trajectories of home runs hit identically—with the ball having the same initial velocity and spin—at an angle of 35° at sea level, in Denver, and in Mexico City. The position of the ball 3, 4, and 5 seconds after it is hit is shown also.*

the playing field is reduced as a consequence of the protection afforded by the stands, the long, high home runs can still be carried along by a following breeze.

Since the retarding force on the ball is proportional to the density of the air, the ball will travel farther in parks at a high altitude. A 400-foot drive by Sammy Sosa or Mark McGwire at Shea Stadium, near sea level, on a windless summer day would translate to a 404-foot drive in Atlanta on the Georgia Piedmont, at 1050 feet the highest park in the majors before 1994. The same home run could be expected to go about 403 feet in Kansas City and 403 feet at the Metrodome in Minneapolis or Wrigley Field in Chicago. These differences are not so great as to modify the game, but Sosa could expect his long drive to travel about 420 feet at mile-high Denver. And if the major leagues are further internationalized someday . . . say to Mexico City, at 7,800 feet, Sosa's blow could sail nearly 430 feet. Figure 2.6 shows the trajectories of long home runs hit with an initial velocity of 110 mph at an angle of 35° at sea level, in Denver, and in Mexico City. Old home run records will be swept away unless the fences are moved out in the high parks.

But even if the fences are adjusted, the high-altitude stadiums will still be a batter's boon and a pitcher's bane. With fences moved back, there will be acres of outfield for balls to fall into for base hits, and, though the pitcher's fastball will be about six inches quicker in Denver, the curve will bite about 20 percent less, which is more important.

With less drag, the ball will also get to the outfielders faster in Denver than at Fenway Park in Boston. Players for the Colorado Rockies have noted that in Denver's outfield, "Fly balls come at you faster and sail farther than you might expect." Indeed, a hard-hit "gapper" between the outfielders will reach the 300-foot mark about two-tenths of a second faster in Denver than at sea level, cutting down the pursuit range of an outfielder by five or six feet—not inconsiderable in this game of inches. Even the range of a shortstop covering a line drive or one-hopper will be cut by about a foot in Denver.

The use of a less lively, "high-altitude" ball would reduce the altitude effect, just as special less lively, "high altitude" balls are used in tennis, though for somewhat different reasons.

Temperature, barometric pressure, and humidity also affect the flight of a ball. The canonical 400-foot home run will go about three feet farther for every one-inch reduction in the barometer and as much as ten feet farther on a 95° July day in Milwaukee than on a 45° April day.[f] The effect of temperature differences on the elasticity of the ball will also have an effect on the distance a batted ball travels.

Humidity per se has little effect on the flight of the ball. Indeed, since water vapor is a little lighter than air, if all other factors are the same, a ball will travel *farther* if it is exceptionally humid, though only by a few inches. The general belief that balls do not travel as far if the humidity is high probably stems from people's experience of windless humid nights when the tempera-ture has dropped from its daytime highs. Then, with the cooler evening air a little denser and no breeze to carry the ball, the ball that would have cleared the fence for a home run in the hot after-noon carries only to the warning track.

Humidity, however, certainly affects the weight and elasticity of balls in storage. Balls stored under conditions of high humidity will gain some weight through the absorption of water from the air, and their elasticity (the coefficient of restitution, discussed in Chapter 5) will be reduced.[g]

Since the retarding force is proportional to the cross-sectional area of the ball, small balls will go farther than large balls. According to the *Official Baseball Rules*, a ball may be as large as 9¼ inches in circumference and as small as 9 inches, as light as 5 ounces and as heavy as 5¼ ounces. The stroke that propels a larger ball 400 feet will drive a smaller ball—of the same weight and elasticity—perhaps six feet farther. The stroke that drives the lighter ball 400 feet will drive the heavier ball—of the same size and elasticity—the same distance, give or take a foot or two. Though the heavier ball will come off the bat a little slower, its greater sectional density will carry it better through the drag of the air and just about compensate for the deficit in initial velocity.

TECHNICAL NOTES

a) The Reynolds number, Re, is the ratio of inertial and viscous forces. The value of the number for a sphere of diameter, r, moving with a velocity, V, through a fluid of density, ρ, and viscosity, μ, is:

$$Re = \rho \frac{Vr}{\mu}$$

For a baseball, $Re \approx 2200\ V$, where the V is measured in mph.

For a given ball velocity, Re is smaller at altitude even as the density is reduced. The value of Re is also reduced with increasing temperature, as that change decreases the air density, which is proportional to the absolute temperature (the temperature above absolute zero, about −460° F). Such an increase in temperature also reduces Re by increasing the viscosity of the air, μ, which is proportional to the absolute temperature of the air—though independent of the air density.

b) The relation for the drag force, F_d, can be written as:

$$F_d = C_d A \rho \frac{V^2}{2}$$

Here $A = \pi r^2$, with r the radius of the sphere, is the cross-sectional area of the sphere, ρ is the density of the air, V is the velocity of the ball, and C_d is the drag coefficient. For $C_d = 2$, this is just the force required to move a column of air the size of the ball to match the velocity of the ball.

If the air passing by an object is divided into small packets — with each perhaps marked by a smoke particle — and the packets flow about the object in paths marked by smooth lines that do not change their position relative to the ball over time as the flow proceeds, we call the paths "stream-lines." If the air were to flow in completely smooth stream-lines around the ball, generating neither trailing vortices nor turbulence (which do change over time), after the ball has passed, the elements of air in back of it would be just as they were before it passed. Hence no energy would have been transferred to the fluid and the drag would be zero — and thus the adjective "streamlined" to describe airplane and automobile shapes that reduce aerodynamic drag. But a baseball is not very well streamlined.

c) For velocities such that the drag coefficient, C_d, does not vary strongly with the velocity, V, of the ball through the air, the Magnus force, F_m, can be expressed as:

$$F_m = K f V C_d$$

where the force, F_m, is measured in pounds-force, the ball velocity, V, is expressed in mph, and the spin frequency, f, is measured in rpm. The measurements of Watts and Ferrar suggest that $K \approx 2 \cdot 10^{-6}$.

Almost all of fluid dynamics follows from a differential equation called the Navier-Stokes equation. But this general equation has not, in practice, led to solutions of real problems of any complexity. In this sense, the curve of a baseball is not understood;

the Navier-Stokes equation applied to a baseball has not been solved. Professor Robert Romer, long-time editor of the *American Journal of Physics*, told me of an eminent physicist who said, "There are two unsolved problems which interest me deeply. The first is the unified field theory [which describes the basic structure and formation of the universe]; the second is why does a baseball curve? I believe that, in my lifetime, we may solve the first, but I despair of the second."

Therefore, the foregoing equation used to describe the Magnus force follows from the fundamental Navier-Stokes equation only after that equation is simplified through some rather drastic approximations. My simple Newtonian description of the complex processes that govern the curveball does not contain all of the truth. But it is *useful*, and surely reasonably accurate for baseball velocities under 75 mph. In the absence of good measurements at higher velocities, we can be less certain of my estimate of the Magnus forces for balls traveling at greater speeds. In particular, the lack of increase with velocity in the drag force at baseball velocities near 90 mph has not been demonstrated experimentally, and the variations of the drag and Magnus forces with the orientation of the axis of the ball, important in this game, are not well known.

Incidentally, the Magnus effect responsible for the curve of the curveball and the hop of the fastball is not quite the same as the Bernoulli effect; it is, indeed, *more* than the Bernoulli effect—which is why it is called the Magnus effect and not the Bernoulli effect.

d) To consider the Magnus force at velocities greater than 75 mph, we must take into account the variation of the drag coefficient. In general, that coefficient, C_d, varies strongly with velocity near the transition between smooth and turbulent flow, and we might reasonably expect the Magnus force, F_m, to vary as:

$$F_m = KfVC_d[1 + 0.5 \cdot \left(\frac{dC_d/C_d}{dV/V} \right)]$$

This differential form follows naturally from the basic equations for both the low-velocity smooth flow and the high-velocity turbulent flow. According to this relation, the Magnus force can reverse sign when the logarithmic differential $(dC_d/C_d)/(dV/V)$ is less than −2, which can occur at the drag crisis when the drag force decreases sharply as velocity increases. This description is not soundly based on theory, but it accounts for the reversed Magnus force seen for smooth balls.

The failure of the Magnus force to increase with velocity for baseballs at velocities from 70 to 100 mph that follows from these arguments, and from the foregoing formula (as shown in Figure 2.2), is the same effect that reverses the sign of the force for balls with more uniform surfaces. This reversal was observed by Briggs in wind tunnel measurements of the Magnus force on smooth spinning balls the size of a baseball. For golf balls with the dimples smoothed out, that dip is sufficiently great as to reverse the Magnus effect, so that a ball hit with backspin ducks into the ground rather than rising, as do normally dimpled balls.

e) Even as the interaction of the spinning ball with the airstream causes the ball to be deflected, there must be a reaction on the ball that reduces its spin rate. The unbalanced drag forces shown in Figure 2.3 that generate the Magnus effect must also produce a countervailing torque, slowing down the spin.

In the wind tunnel system of Figure 2.3, the interaction of the surface of the spinning ball with the air tends to slow down the air more on the side of the ball moving faster through the air (the third-base side in Figure 2.3) and less on the other side (the first-base side in the figure). This differential change in the velocity of the air results in a countervailing change in the pressure through the Bernoulli effect—the faster the flow, the lower the pressure—which then contributes in a major way to the Magnus force. But those changes in the velocity—and thus momentum—of the air that differ on the two sides of the ball require a tangential force that must be supplied by the moving ball surface.

Hence, we can expect a reactive torque, L, that resists the spin, such that:

$$L = kF_m r$$

where F_m is the Magnus force, r is the radius of the ball, and k is a proportionality constant that I estimate as $\approx 1/10$. (I note that the estimate is not solidly based.) The energy-per-second or power, P, lost to the spin kinetic energy, T, is then:

$$P = L\omega = kF_m r\omega$$

where $\omega = 2\pi f$ is the angular spin velocity of the ball.

The spin-down time constant, τ, is then:

$$\tau = \frac{T}{P} \ where \ T = \tfrac{1}{5}mr^2\omega^2$$

where m is the mass of the ball.

The time constant, τ, is a measure of the time required for the ball to lose its spin. Thus, for, $\tau = 5$ seconds, a value that I estimate to be typical of normally spinning fly balls, the spin rate will decrease by a factor very near $1/\tau = 1/5$ for each second the ball is in flight.

At high spin rates, the drag force on a baseball may increase. Some wind tunnel measurements show substantial increases in drag for spinning balls; some do not. In science, unlike many other disciplines, if two results differ widely, the truth is more likely to be found at one of the extremes than in the middle. However, my experience with the game of baseball itself leads me to believe that the drag on a baseball does not increase strongly with spin.

For moderate spins, where $v'/v < \frac{1}{2}$, I assume that the drag is increased by a factor of $1 + (v'/v)^2$, where $v' = \omega r$ is the rotational velocity of the surface of the ball and v is the ball velocity. This form fits some measurements and has a theoretical basis. For larger values of v'/v I use a more arbitrary recipe, but that only affects pop flies and foul balls, where there are other uncertainties as well.

f) For baseballs moving at very high velocities—over 120 mph—or at low velocities—less than 60 mph—the drag coefficient, C_d, does not vary much with velocity. At these velocities the drag resistance on a baseball is simply proportional to the density of the air and hence, for a given temperature, to the barometric pressure. An increase in altitude of about 275 feet (equal to a reduction in pressure of a third of an inch of mercury on the barometer) or an increase in temperature of about 5° F decreases the drag by about 1 percent.

But the situation is more complicated for the intermediate velocities that are most important to baseball. At these velocities, where the transition is taking place from high-drag smooth flow to low-drag turbulent flow conditions, the drag coefficient is changing from values of about 0.5 at low velocities to values of about 0.25 at high velocities. Since the drag coefficient is known to be largely dependent on the Reynolds number, which depends on the pressure and temperature as well as velocity, the "position" of the transition as seen on Figure 2.1 moves to higher velocities as temperature and altitude increase. Thus this effect, acting alone, *increases* the drag on a baseball moving at the ordinary velocities of the sport as either the temperature or altitude increases, partially countering the reduction in drag caused by the decreased density of air being pushed aside by the ball.

We do not know the precise variation of the drag coefficient with velocity well enough to reliably calculate the effects of that variation on the drag force, but I estimate (that is, I make an educated guess) that the variation in the transition with temperature and altitude reduces the density effects by about a factor of two. With this factor included, the drag is reduced by 1 percent for each increase in altitude of 550 feet and for each increase in temperature of 10° F.

For the 400-foot home run, this 1 percent decrease in drag by 1 will add a little less than two feet to the ball's carry and increase the probability of the average player's hitting a home run by roughly 3.5 percent.

When all is considered I find that, as a consequence of the reduced air density at high altitudes, the drive projected at an angle of 35° upward at sea level that lands 400 feet from home plate will travel an extra 4 feet for every thousand feet of altitude where conditions are otherwise the same. A 350-foot drive will gain an extra 4.6 feet per thousand feet of altitude. The approximation that the increase in distance is proportional to the increase in altitude is probably good to a few percent.

The problem of the smooth-turbulent transition variation does not occur for golf balls, where the dimples place the ball in the turbulent region for most velocities important in golf. Thus the drag on a golf ball is simply proportional to the air density. With that simplification, I find that a golf ball will fly about 11 percent farther in Denver than at sea level, a conclusion confirmed by professional golfers.

g) R. C. Larsen found that the weight of balls stored at 100 percent humidity for four weeks increased by 11 percent and that the coefficient of restitution at an impact velocity of 25 mph decreased by 10 percent—when dropped on concrete from a height of 20 feet, the humidified balls will bounce only about 80 percent as high as the balls stored at low humidity. If that proportional decrease in elasticity would hold at greater-impact velocities, the swing of the bat that would drive a "dry" ball 380 feet would propel the ball stored at high humidity only 350 feet. After sitting in dry air for a few hours to dry their covers, the humidified balls could not be distinguished from normal balls by a layman. However, an experienced pitcher would have probably noticed that they were heavier and softer than the balls he was used to.

THE SWING OF THE BAT

A MODEL OF BATTING

There are many ways of batting the ball successfully through swinging at the ball in a manner such that precision of placement of the batted ball is important rather than high velocity. The drag bunt and the hit-and-run behind the runner at first base, for instance, represent purposeful hits that are made by swinging with less than full power. Hits to the opposite field often fall into the same category. Also, the batter may be fooled by the pitcher to such a degree that his timing is thrown off and he swings weakly—but sometimes successfully, hitting, perhaps, a "Texas leaguer" or an infield grounder which bounds so slowly that he beats out the throw to first. In the following discussion, I don't consider the imperfect swings—though they may constitute the majority—but only the full swing made with maximum effort. Just a few players—Babe Ruth and Ted Williams come to mind—*always* seemed to make full swings.

In considering the mechanics of the batting process, it is useful to construct a model of a full swing that is tractable but sufficiently close to a real swing by a real batter that the consequences of the simple model illuminate the complex reality. Such modeling is simplified by the realization that a player swings a bat very much as he would a weight on the end of a rope; to a very good approximation, the forces on the bat exerted through the hands are directed along the axis of the bat. Hence, if the motion of the hands is known throughout the swing, the motion of the bat can be determined.

Such a model swing at a waist-high pitch by a strong player is

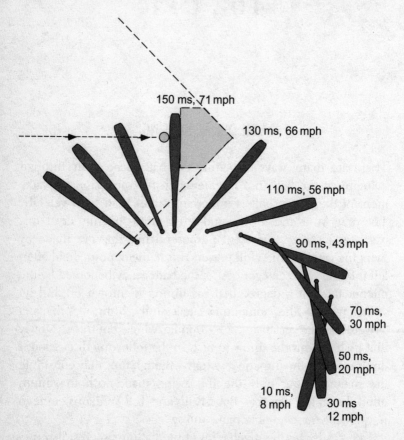

FIGURE 3.1: *The trajectory of the bat during a typical swing. The position of the bat is shown at intervals of ⅟₅₀ of a second. The player will actually begin the swing about 20 or 30 milliseconds earlier, but the bat moves very little at first and the batter is not committed; hence, I plot only the last 150 milliseconds.*

described in Figure 3.1, which shows a plausible trajectory of the bat throughout a swing in a horizontal plane, calculated from simple dynamics and the trajectory of the batter's hands. The bat velocity is taken as the velocity of the center of gyration with respect to the end of the bat, which is near the vibrational node, or "sweet spot," of the bat—about the best place to hit the ball. In the figure—and the calculation—the initial position of the bat is taken as lying in the plane of the swing for convenience, though most batters actually hold the bat more nearly upright. That difference does not seriously affect the consequent bat trajectory.

If the batter is to swing at the ball, he must begin about 30 milliseconds earlier than shown in the picture, but those actions involve no commitment, hence I show just the 150 milliseconds before the bat strikes the ball—if it does.

The graphs in Figure 3.2 show the variations with time during the swing of the velocity of the hands, the velocity of the barrel of the bat, the total force on the bat and the component of force in the direction of motion of the hands, the power supplied to the bat, and the kinetic energy of the bat.

Before the pitch, the batter usually stands with his weight placed over a point about halfway between his feet, which are spread to about shoulder width; typically, the back foot is planted near the rear edge of the batter's box. When the ball approaches the plate, the batter generally shifts his weight backward so that it is centered near the back foot. This motion is halted and reversed by a large push from the back foot as the batter moves his body toward the pitcher, usually in the course of a short step.* During this "stepping into the pitch," the strong, 180-pound batter pushes off his rear foot with a force of perhaps 250 pounds and, over a period of about 0.2 seconds, reaches a velocity of about 6 mph. During the first part of this step—the first 0.15 seconds of the swing—while his front foot is in the air, the batter hardly rotates his body and rotates the bat only moderately. His body is

---

*Babe Ruth often took two steps, beginning the first one about the time the ball left the pitcher's hand.

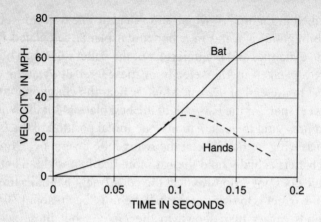

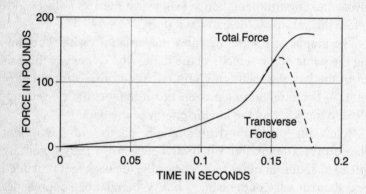

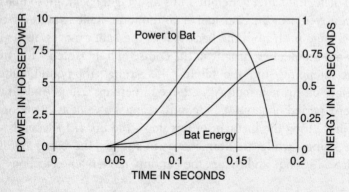

FIGURE 3.2: *The variation with time of the velocities (top), forces (center), and power to the bat and bat energy (bottom) during the model swing.*

still cocked and he is looking over his shoulder at the incoming ball. Only after about 0.05 seconds, when the batter has only moved perhaps an inch, does he begin to bring the bat around.

When his front foot is securely planted, he begins to rotate his body around that foot, which is held rigidly and used as a pivot. The front foot is placed a little behind the initial line of motion—which is usually slightly toward the plate—so the forward momentum is turned to rotation by the reaction of the stiff front leg. Also, significant rotational energy is then generated as the muscles of the torso act through torque from the feet, both of which are now firmly placed on the ground. Then, in the next tenth of a second, the batter rotates his body—hips, torso, and shoulders—as he brings the bat across the plate. Most of the translational and rotational energy, generated in the complex actions over a fifth of a second that bring the body into play, is now concentrated in the bat.

Typically, the center of mass of the batter moves forward a little less than a foot in the fifth-second required to swing the bat. The considerable energy of about 0.6 horsepower-seconds (a horsepower-second is equal to the energy output of a 1-horsepower motor running for a second, enough to lift 550 pounds one foot) is transferred to the bat in that time (the *average* rate of transfer of the energy is then about 3 horsepower). That energy is generated largely by the large muscles of the thighs and torso. The arms and hands serve mainly to transfer the energy of the body's rotational and transverse motion to the bat and add little extra energy to the bat.*

We have described the swing in two parts, an initial step-in part moving the batter forward, and a final rotational part—

---

*In particular, the contribution of the hands and wrists to the energy of the bat is almost negligible. Though the long-ball hitters—generally big, strong men—usually have large, strong hands and wrists, there are exceptions. For the first century of its existence, no National League player hit more home runs in a season than Hack Wilson, 5'6" tall and 190 pounds. Wilson, however, built like a fireplug, wore size 5½ shoes, and his small hands, "which matched his feet," could not have been especially strong. But the tremendous strength in his legs and torso was transferred efficiently to the ball through his strong arms.

though the rotation starts while the body is still moving forward. Those two parts roughly coincide with bat action and reaction phases. For the first tenth of a second, while the batter is striding toward the pitcher, the character of the swing is dominated by the action of the batter; for the second tenth, until the ball is struck, the character of the swing is largely determined by the reaction of the bat.

During that first part of the swing, the hands transmit an increasing force that reaches nearly 50 pounds, largely generated by the rotation of the body in pulling the bat about an arc. At the end of that time, the hands and the trailing bat reach a velocity of about 20 mph and the kinetic energy of the bat is about 0.05 horsepower-seconds. The weight shift during this time is equivalent to a motion of the body toward the pitcher with a velocity of about 6 mph corresponding to a kinetic energy of perhaps 0.4 horsepower-seconds for a 180-pound batter. More energy, perhaps 0.2 horsepower-seconds, is stored in the rotation. Thus about 0.6 horsepower-seconds (330 foot-pounds) energy is stored in the motion of the body.

After one-tenth of a second elapses, the bat moves out of the arc of the hands and can be considered to exert an increasingly large reaction force on the hands and arms that will reach a value near 200 pounds as the bat crosses the plate. During this time, that large force straightens (uncocks) the wrists. The force from the reaction of the bat, transmitted through the hands and arms, slows down the motion of the body, which is now mainly rotary. In the sweet swings of the great batters, the stored kinetic energy of the body is transferred efficiently to the bat,* increasing the velocity of the bat from 20 mph to about 70 mph. The rate of that energy transfer exceeds 8 horsepower about 0.14 seconds after the start of the swing—and about 0.04 seconds before the ball is

---

*The part of the total force of reaction by the bat in the direction of motion of the hands is responsible for the energy transfer. That force ranges up to 150 pounds, requiring strong arms. This is why Pete Gray, the one-armed outfielder who played for the St. Louis Browns in 1944, hit no major league home runs. Using only one arm, he could not apply a force as great as the 150 pounds necessary to transfer enough of the energy developed in his strong thorax and legs to the bat to enable him to hit the ball a great distance.

hit. This 8 horsepower is far greater than that actually generated by the contraction of muscles during this period. For many batters, the energy transfer from body to bat is nearly complete—the forward motion and rotation of the body is stopped almost completely when the bat crosses the plate.

The discussion above was centered on energy transfer. A complete analysis would also include momentum transfer and the consequent recoil of the upper body of the batter. For some batters, the recoil is quite noticeable; as the bat goes forward to strike the ball, the head and shoulders recoil backward.

There has been some controversy over the relative importance in batting of rotational motion and translational motion, Ted Williams emphasized the importance of rotation, and Charlie Lau emphasized translation. Of course, both are essential (as both Williams and Lau knew), and they are interrelated even as the energy of translation of the body goes into the rotary energy of the bat. Those who consider translation unimportant should imagine a batter standing with his back foot on a platform set on ball-bearing wheels on a toy train track running parallel to a line from the pitcher to the plate and with another such platform to catch his front foot—presuming he was able to place it there. This batter would not be able to stride forward, though he could still rotate. But how far could he hit a ball? I say not much past second base.

On the other hand, if the tracks ran perpendicular to the line to the pitcher-plate line, he would be able to stride forward but would not be able to rotate. Without the possibility of a strong rotation, he would, again, be lucky to clear second base. To hit a baseball with dispatch, one needs to step into the ball, and to rotate.

Typically, the fastball struck by the bat carries about a fifth as much kinetic energy as the bat. If the ball is struck squarely, about half the energy of the swinging bat is transferred to the ball in the impact; hence, the speed of the bat is sharply reduced (by about 30 percent) by the collision. About one-third of the original bat-and-ball energy is carried off as kinetic energy in the flight of the ball from the bat and the rest of it (about three-tenths of the original total energy) is lost in friction in the course of the distortion of the ball—and then in heating of the ball. (Indeed, if you catch a Mark McGwire blast in the bleachers, you might be able to sense the side of the ball that he hit from its slightly increased temperature.)

The course of the swing after the ball is struck is kinematically irrelevant. The emphasis by coaches and sports teachers on the "follow-through" in baseball (and in golf and tennis, among other sports) is designed to ensure proper actions *before* the ball is struck. Figure 3.1, however, does show typical positions of the bat and the hands after the ball is struck; the pattern would be a little different if the ball were missed.

The graphs in Figure 3.2 show a typical variation of the bat and hand velocities during the swing, the variation of the forces applied to the bat, and the energy transfers during this model swing. Though the precise quantities shown in the graphs derive from this particular model, the qualitative values of the quantities and the character of the changes of the quantities that are presented are probably close to those of real swings.

The batter in the diagram is assumed to have swung in such a manner as to drive a ball to center field as far as he can. If this right-handed batter* should miscalculate the velocity of the pitch and hit the ball five milliseconds late, and to right field, he would hit the ball before maximum bat velocity was realized and not generally hit it quite as hard. Since the loss of energy after the maximum is reached is usually small, if he swings five millisec-

---

*Though the right leg bears more of a burden than the left for a batter who swings from the right side, the arms share the forces nearly equally. Hence, there is no natural advantage for a right-handed man to bat right-handed. Indeed, many players who throw right bat left, since the left-handed batter is favored in many ways: The batter who swings from the left side is closer to first base and is already moving in that direction as he finishes his swing; for some reason right-field fences tend to be closer than left-field fences; and there are more right-handed pitchers (whose slants are easier for a left-handed batter to hit) than left-handed pitchers. However, there are two famous first basemen who threw left and batted right: Hal Chase and George H. Bush. Chase, a great-fielding first baseman, was considered the best ever at the position by such as Walter Johnson and Babe Ruth, while Bush, captain of a fine Yale team that went to the finals of the NCAA tournament, did an exemplary job for a few years of throwing out the first ball on opening day.

onds too early, he will lose little power in driving the ball to left field.

A pull hitter swings so as to maximize the bat velocity a little later in the swing. In general, the pull hitter has more time and distance to apply force to the bat and hence can transfer a little more energy to it and hit the ball a little harder. If the right-handed batter makes a minimal adjustment in the position of the arc of his swing, he will naturally hit a ball over the inside portion of the plate later in the swing, and to left field. He will hit a ball on the outside of the plate earlier in the swing, and to right field. It is then harder to hit a very fast ball over the inside of the plate, and "high and tight" is a natural strikeout pitch for an exceptionally fast pitcher.

Aiming for the center-field bleachers, the batter in Figure 3.1 will be swinging the bat at a speed (of its prime hitting region) of approximately 70 mph, in a manner such that the bat, as it passes through the hitting zone, is effectively rotating about a point near the handle end of the bat. If he swings as much as $\frac{1}{100}$ of a second early, the ball will go foul down the left-field line; if he is $\frac{1}{100}$ of a second late, the ball will go foul into the stands down the right-field line.

Batting is an intricate art, and there are various ways to bat well. If the batter in Figure 3.1 swings in such a manner (more of an "arm" swing) that his hands move faster through the hitting zone such that the bat is effectively moving in a longer arc, perhaps as if it were rotating about a point about half a bat length beyond the end (toward the player), the bat will be rotating slower at the time of impact and the error of $\frac{1}{100}$ of a second, early or late, would result in a long fair fly to left or right field rather than a foul ball. With such a swing, less precision in timing is required. However, the longer motion must be started a little sooner than the more compact arc of the "wrist" hitter.

Often tall, strong hitters, like outfielders Darryl Strawberry and Dave Winfield ca. 1990, use a longer sweep and longer arc, balancing the fact that they have to initiate their swing sooner (with more chance of misjudging the pitch) with the greater power generated in a longer swing and by the less precise timing required in a sweeping swing. They also can afford to use a heavier bat. Smaller players who hit with power, such as Howard Johnson, who played about the same time, more often tend to rely on quick, precisely timed swings with shorter arcs, such as shown in Figure 3.1. They will choose a lighter bat to help them swing more quickly. With such swings, they can delay their commitment and better judge the ball.

Most distance hitters hold the bat at the end to maximize the bat velocity realized from the whip motion. Babe Ruth can be considered to have held the bat beyond its end! Batting left-handed, for much of his career he held the knob of the bat in the palm of his right hand.*

Players of yesteryear who emphasized precision—like Wee Willie Keeler, who "hit 'em where they ain't"—often used long arcs and reduced the angular velocity. Ty Cobb and Honus Wagner both separated their hands on the bat and swept the bat through the hitting zone, adding to the angular accuracy of their hitting (though Cobb, especially, would often bring his hands together early in the swing).

BATTER REACTION TIME

In the course of batting, the hitter observes the pitcher closely, sees the ball delivered, and makes a judgment concerning the probable ball trajectory that leads to a decision to swing or not to swing at the pitch. And if the batter decides to swing, he must further decide as to just when and where and how he will swing so that the hitting area of the bat strikes the ball squarely—he hopes! All of this takes time and judgment—and *good* judgment takes more time. The pitcher must work to upset that judgment.

---

*Ruth, like most players, often made minor changes in his batting style. Pictures of the Babe on his follow-through after hitting his 60th home run off Tom Zachary in 1927 clearly show the knob of the bat below his right hand.

He may do that by throwing the ball so fast that the batter has too little time to evaluate the pitch properly. He may do that by throwing in an unusual trajectory—perhaps an especially sharp-breaking curve—so that the batter cannot properly judge that trajectory. He may do that by disguising the pitch so that the batter picks the wrong solution and swings at a "fastball" that turns out to be a forkball. He may throw an occasional brush-back pitch so as to require the batter to make an extra decision on inside pitches—to hit the deck or not. Thus an analysis of the processes that convert a batter's vision of the ball into a decision to swing at that ball might give us some special insight into the pitcher-batter duel that lies at the center of baseball.

Some of this insight can be gained by considering the time required for a very simple reaction—a *voluntary* "blink of the eye." The eye takes about 50 milliseconds to blink involuntarily after a jet of air is blown into it. But a voluntary blink, done by design after one sees a light flash, takes about 150 milliseconds. Thus, the minimum time for a deliberate muscular response to a visual signal is about 150 milliseconds.

There is a considerable advantage for sprinters running the shortest distances in track to get a rolling start. The runner who, properly, only begins his run once he sees the smoke from the starter's pistol will not actually push off from the starting blocks until about 200 milliseconds later. He will need the 150-millisecond minimum reaction time to see the smoke (or hear the sound), process that information, and then send signals from the brain to the muscles that will push him off the blocks. Then it will take a little more time for the muscles to react, and still more time to accelerate him across the starting line. Thus the runner who guesses when the gun will go off and starts 200 milliseconds early will gain about two yards on his by-the-book companion without crossing the starting line too soon for an obvious false start. To stop this guessing, and the many false starts that result from bad guesses, starting blocks in major track meets are now instrumented to record push-off times, and the athlete who pushes off sooner than 100 milliseconds after the gun is fired is charged with a false start.

With this minimum time of 150 milliseconds required to blink an eye or to start the 100-yard dash in mind, we can consider the problem of a man who is supposed to swing a two-pound round piece of lumber in a big arc to strike squarely a ball that is directed toward him from sixty feet away at a speed that brings that ball to him in 400 or 500 milliseconds. The batter then has just two or three blinks of the eye to see the pitch, judge the correct response, and make the movements required to hit the ball squarely. We can see that he is faced with a very difficult problem. But professional ballplayers somehow solve the problem, and the best do it remarkably well.

Our understanding of the mental processes that take place in the brain is still in its infancy, but with the knowledge we do have, together with our insights as to how information is handled, we can make some useful deductions concerning the way the batter takes in and processes the information that leads to the swing of the bat that propels a fastball into the bleachers—sometimes. And with that insight, we can better understand the pitcher's efforts to foil the batter.

Let us apply this perspective to the batter's information processing during the time of flight of the pitched ball.

Looking: As the pitcher's hand travels toward the plate just before he releases his pitch, the batter is watching intently to get a first insight into the possible spin and trajectory of the ball. Light from the pitcher and from his background is focused by the lenses of the batter's eyes onto a mosaic of about a hundred million sensory cells lining the retina at the back of the eyeballs. Those cells, mostly rod-shaped in the periphery, wholly cone-shaped at the center, generate electrical signals that are initiated by light-absorbing chemical processes in the cell. These signals are processed by sets of cells in the retina that encode the information in further sets of tenth-volt action pulses that are transmitted to the brain by the bundle of about a million axon nerve fibers that make up the optic nerve. The encoded information is not arranged simply as a set of pixels but more as sets of borders of contrasting areas, together with change or motion information. The assembly of the information by the eye and its associated nerve system and the generation of the electrical pulses that encode this information seem to take a total of about 25 milliseconds.

While signal pulses travel in copper wires at speeds near the velocity of light, the signals in the nervous system travel about ten million times slower. The largest of these physiological wires, extending along the spinal cord to the major muscle groups in the limbs, are about $\frac{1}{1000}$ of an inch in diameter and carry signal pulses at a speeds near 200 mph—about twice the speed of the fastball. The signals travel more slowly in the smaller axons that make up the optical nerve and the pulses carrying the information from the eye take about 20 milliseconds to reach the back of the brain, where the visual information, not yet a useful picture, is processed by the brain-as-computer.

The brain-computer is very slow—a switch opening or closing takes at least a millisecond, while your PC switches about a million times faster. But the brain-computer has about a hundred trillion connections (synapses) while your PC gets by with about a millionth of that number. Thus the brain can be considered to act more like ten million individually powerful but very slow ten-million-connection computers, working simultaneously—or in parallel.

Although a serial computer like your PC would take hours to assemble a picture based on the information from the eye, the brain-computer, with thousands of subcomputers working simultaneously in parallel, takes only about 30 milliseconds for the construction. Some elements of the constructed picture were not even present in the information from the eye but were supplied from patterns stored in the brain-computer memory. For example, the eye, focused as it was on the pitcher, did not record the green of the infield grass behind first and third base, but the "green" is filled in by the brain with information stored from the batter's look at the grass a second or so before, when he surveyed the positions of the infielders.

Thus about 75 milliseconds, the time a fastball travels about nine feet, elapses from the time of the first observation by the batter through the construction of the picture.

After the picture of the ball leaving the pitcher's hand is processed, the eye continues to record information. This information is transmitted more nearly continuously than it is for individual snapshots, but the 25-millisecond eye-processing period is still relevant. Hence, for the next 400 milliseconds, while the fastball sails from pitcher to plate, the eye transmits information on the ball position that is not very different from that contained in something like sixteen frames of a motion picture.

Thinking: When the brain does engage in analysis with the information about the ball's path to the plate, does it then calculate the ball's trajectory? Well, in the way your PC, properly programmed, would do the calculation;[a] the brain is much too slow and the information—especially that concerning the timing—is not good enough. But it does compare the pattern of information it has received on the position and motion of the ball with patterns it has stored from the ballplayer's long experience. The patterns "on top" of the stack of patterns stored in the brain have been preselected to some extent by the batter's thoughts before the pitch was thrown. If the pitcher is a fastballer, the review of the situation the batter conducted while he was in the on-deck circle serves to make a special subset of patterns most easily available. Another set will have been shuffled to the front of the deck if the batter judges the pitcher as a junk pitcher likely to throw soft, off-speed pitches. Somehow, these patterns are scored and the brain chooses the best one. Perhaps, from the many patterns labeled (in brain-computer language that we might call "mentalese") "fastball, curve, or slider; high or low; in or out," the brain will select "high inside fastball for a strike from an over-the-top pitcher" as best matching the input information.

The longer the brain has to think, the more accurate and more detailed will be its choice, but if the batter is to swing at the fastball in time to hit it, it seems that he will have only about 25 milliseconds to select the pattern that fits the pitch.

Having "decided" the character of the pitch, the brain-computer must select the proper swing pattern (which includes a no-swing choice and, maybe, a "hit-the-deck" pattern) from quite another set of memories that seem to be stored in another part of the brain. Those swing patterns have been established and refined in the course of his long experience in games and practice and cover about every pitch he has ever seen. Again, the choices are preselected to some extent by his assessment of the strategic pattern of the game, the ball-strike count, the character of the previous pitches, and perhaps the instructions of his manager or coaches. With his choice of the pitch pattern made, he must select a swing pattern that matches; perhaps "a level swing at a high, inside fastball, designed to meet the ball in a line-drive mode" (rather than an upward swing designed to send the ball over the fence). That pattern comes as a set of commands from this "planning" region of the brain, which sends with exquisite timing over a period of more than 100 milliseconds hundreds of tenth-volt signal pulses out through hundreds of nerve axons. These pulses go to different regions of the brain, separated by inches, which direct similar tenth-volt signals through the nerve axons to the player's right leg muscles, left leg muscles, right arm muscles, and so forth, generating the beautiful swing of the professional ballplayer.

Again, the longer the time available, the more detailed the selection process and the higher the probability that the best pattern will be chosen, but the fastball hurries the choice, which must be made in about 25 milliseconds if the batter is to complete the swing in time to hit the ball.

Action: Once the initial "looking" and "thinking" have been completed (in about 125 milliseconds), the signals are only just being sent out by the brain and must still travel to the muscles. Perhaps the first muscle group to be activated is that of the back foot and leg. The primary "wires," or axons, that transmit that signal from the brain to the leg are large and the signal travels quickly. About 15 milliseconds are required for the fastest signal to get from the brain to the vicinity of the lower leg, a little less to get to the thigh. Another 10 milliseconds are required to distribute the signals to the various muscle fibers through smaller and slower distribution axon-wires, each of which activates a set of from ten to a thousand muscle cells. For each muscle that contracts to move a limb one way—perhaps to straighten the knee in order to extend the leg—other muscles that move the limb in the opposite way—to bend the knee—must receive signals telling them to relax. Other muscle group pairs are then activated in turn as the swing progresses, all from signals produced almost automatically based on the pattern selected by the brain and sent out at carefully designed times over the 150 milliseconds that it takes to swing the bat.

Batting: After the signal from the brain reaches the muscles through the nerve-cell axons, the muscle action takes time to develop; a short fast-muscle cell may be completely activated in as short a time as 10 milliseconds, whereas a longer fast-muscle cell may take as long as 50 milliseconds to contract completely. Thus bat-swing time is not simply that required to move a mass under a force, but is determined also by the speed of the biochemistry involved in the flexure of the muscle cells.[b] It takes about 180 milliseconds for a full swing of the bat, though there is no commitment for the initial 30 milliseconds, which are occupied largely in transferring the batter's weight from back to front foot and getting the body in motion. Hence, I'll consider only the 150 milliseconds after that period. But the timing of that 150 milliseconds is delicate: If the swing of a right-handed batter is 7 milliseconds too late, the squarely hit ball will sail foul past first base; if it is equally early, the ball will travel on the foul side of third base.

In summary: The whole process just described, from the look at the pitch to the moment the bat crosses the plate, takes about 300 milliseconds. The timings required to begin the swing in time to meet the very good fastball are shown are shown in Figure 3.3 by the heavy arrows. But even though the player has started to swing, he still has some time to adjust the bat trajectory. He can make significant changes in the first 50 milliseconds of the swing. In making these modifications, he uses information he gathers by "looking" at the ball for another 50 milliseconds and then "thinking" and "acting" in response to that look. This period of further information is shown by the shaded areas in the figure.

The exceptional athletes who play the game can probably make small adjustments for another 50 milliseconds (also shown in Figure 3.3), though at the end of that time the bat is traveling at about three-fourths of its final velocity. Even if the skilled batter can use this late data, none of the subsequent information

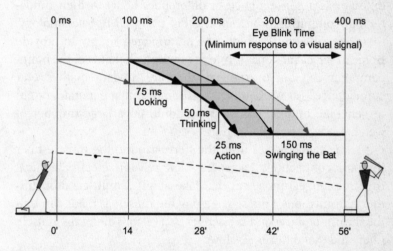

FIGURE 3.3: *The times and distances for a fastball trajectory, showing periods of looking, thinking, and acting. The arrows connecting the different activities show cause-and-effect relations, indicating the times required for the activity. Thus the player cannot begin "thinking" about an observation until 75 milliseconds after the time it occurs. Similarly, the muscles cannot begin an action until 25 milliseconds after the thinking process in the brain orders that action. The heaviest arrows show the optimum initiation times of the activities if the swing is to be successful. The shaded area extending another 50 milliseconds suggests times during which further observations and their associated actions may modify the swing in a useful way, and the last 50-millisecond set indicates times during which observations may allow some slight adjustments of the swing. After that period, no observation can be used to affect the swing.*

from seeing the ball over the last half of its flight can be used at all. If it weren't psychologically upsetting, the batter could just as well close his eyes after the ball is halfway to the plate or, if it was a night game, management could turn out the lights—the batter would hit the ball just as well.

TECHNICAL NOTES

a) The architectures of the PC and brain-computer are quite different as a consequence of differences between their fundamental computing units. For the PC using CMOS technology, these are "inverters" which can be arranged simply to provide Boolean logical actions. While the neuron elements of the brain-computer can be so arranged, they are fundamentally "voter coincidence" (or McCulloch-Pitts) circuits that generate, or fail to generate, output pulses upon adding up a large number of weighted "yes" and "no" inputs.

b) The signal passes from the nerve-axon to the muscle fiber through connections (synapses) that rely on electrochemical (first messenger) processes that take about 2 milliseconds. Further transmission of the signal along the muscle is made through the diffusion of (second messenger) Ca^{++} ions along the muscle fiber, and that diffusion is slow.

PITCHING

THE PITCHER'S TASK

The most important person on the team, in any one game, is the pitcher.* The pitcher's job is simply defined—if not simply executed. From the pitcher's mound, 60'6" from the rear point of home base, he must throw the ball over the plate for strikes. But within that constraint he must project the ball in patterns of trajectories, velocities, and placements such that the batter cannot hit the ball squarely. But what is a strike? Of course, if a batter swings at the ball and misses, he is assessed a strike. But more important,

> A STRIKE is a legal pitch when so called by the umpire, which (b) is not struck at, if any part of the ball pass through any part of the strike zone.
>
> The STRIKE ZONE is that area over home plate the upper limit of which is a horizontal line at the midpoint between

---

*Though over the whole season, the best hitter in baseball—Babe Ruth—was judged to be more valuable than one of the best pitchers in baseball—Babe Ruth. In 1919, Ruth was moved from the Boston Red Sox pitcher's box, where he worked every fourth day, to an everyday job in right field.

the top of the shoulders and the top of the uniform pants and the lower level is a line at the hollow beneath kneecap. The strike zone should be determined from the batter's stance as the batter is prepared to swing at a pitched ball. [*Official Baseball Rules: 2001*, pages 21–22].

1.05 Home base ... [is] 17 inches long ... with the 17-inch edge facing the pitcher's plate [*Official Baseball Rules*, page 5].

Using these definitions, for a typical six-foot batter the top of the strike zone will be about 45 inches above the plate and the bottom about 19 inches above it. With these definitions, a ball not swung at will be called a strike if it passes over the plate so as to nick a five-sided box with a cross section the shape of the plate — 17 inches wide, 26 inches high, and 17 inches deep — that might be suspended with its bottom edge 19 inches above the plate.

As any ballplayer can tell you, in fact the zone is where the umpire says it is, and as a matter of practice at the turn of the millennium, most umpires seem to define the high boundary of the strike zone about 4 inches lower than the rule book says, and most give the pitcher an inch or two more on the outside corner and almost as much inside. Hence the center of the ball must strike a three-dimensional target about 23 inches wide by 26 inches high and 17 inches deep. Control pitchers hit corners with an uncertainty of about 4 inches; one would have to be a fairly good shot to shoot a pistol with that accuracy. A mistake on one pitch of a hundred, such as an error of 12 inches that puts the ball in the center of the strike zone, may lose the ball in the bleachers — and the game.

The pitcher must not only throw the ball so that it passes through the strike zone, he must throw so that the batter does not hit the ball squarely. He does this by throwing the ball so that it passes through specific places at or near the strike zone that the batter finds difficult to reach, by throwing the ball at different velocities to upset the batter's timing, and by applying spin to the ball in such a manner that the ball passes the batter with different trajectories, confusing and confounding him. Thus, the pitcher varies the placement, the velocity, and the movement of the ball. The pitcher's action up to the release of the ball is part of the art of pitching; the action of the ball after release, determined by the laws of nature, is addressed by physics and is subject to our analyses.

THE CURVEBALL

Cooperstown gives precedence to Candy Cummings as the first pitcher to throw a curveball, though Cummings was required by the rules in place in the 1860s to throw underhand, rather as a softball pitcher does today. Freddy Goldsmith and others confounded batters with curveballs at about the same time, near the beginning of the history of serious baseball.

As we have noted, balls curve as a consequence of asymmetries in the resistance of the air through which they pass. If the air resistance is greater on the third-base side than on the first-base side of the ball, the ball will veer toward first. While the deception of the curveball follows both from its reduced speed, as compared with the fastball, and—perhaps more important—its drop or downward curve, we will emphasize here the left-right deviation from which the pitch gets its name.

The diagram at the top of Figure 4.1 shows the left-right trajectory (reconstructed from Briggs's wind tunnel measurements) of a rather wide-breaking curveball thrown so as to rotate counterclockwise—as seen from above—by a right-handed pitcher. This ball is thrown with an initial velocity of 70 mph, spinning at a rate of 1600 rpm, to cross the plate about 0.6 seconds later at a speed of about 63 mph. Although the radius of curvature is nearly constant throughout the ball's flight, the deflection from the original direction increases approximately quadratically with distance, i.e., four times the deflection at twice the distance. Halfway from pitcher to plate, the ball has moved about 3.4 inches from the

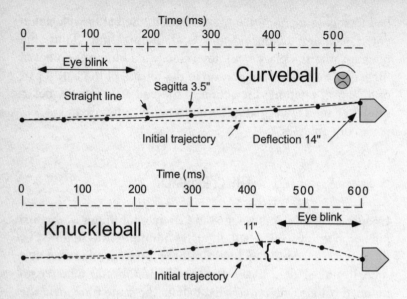

FIGURE 4.1: *The left-right trajectories of a curveball and a knuckleball on their way from a right-handed pitcher to a batter. The curve is rotating counterclockwise as viewed from above the ball's line of flight.*

original line of flight, which is directed toward the inside corner, and is moving toward the center of the plate. At the plate, the deflection is 14.4 inches and the ball passes over the outside corner. From the perspective of the batter—or pitcher—the ball that started toward the inside corner has "curved" 14.4 inches to pass over the outside corner. Moreover, one-half of the deflection occurred during the last fifteen feet of the path to the plate. (I realize that the most useful curve from a tactical view curves down much more than sideways, but I discuss the transverse motion for expositional simplicity.) Does a curveball then travel in a smooth arc like the arc of a circle? Yes. Does the ball "break" as it nears the plate? Yes. Neither the smooth arc nor the break is an illusion; they are different descriptions of the same reality.

Taking the minimum effective information-processing-reaction time (from the discussion in Chapter 3) as 300 milliseconds, we see that the batter must decide to swing when the ball has curved

only about three inches—though he can redirect his swing a little over the next 50 milliseconds. Hence, he has to guess—in
some sense—as to where the ball will go. Some fine athletes
never seem to learn to "guess" the curveball trajectory and are
not, then, found in the major leagues.

Though the deflection of the curve shown in Figure 4.1 as
seen, correctly, by batter and pitcher is 14.4 inches, the sagitta—
the largest deviation from the straight line drawn from the beginning to the end of the ball's flight, as shown in the figure—is but
3.4 inches, only a little greater than the diameter of the ball. A
confusion between the deflection and the four-times-smaller
sagitta has played a role in the canard that a baseball does not
curve. In 1941, *Life* magazine arranged for the photographer
Gjon Mili to use high-speed flash lamps to photograph curveballs thrown by Cy Blanton and curveballs and screwballs thrown
by Carl Hubbell. These fine pitchers threw their major league
curveballs—which drop more than curve—with about 8-inch
left-right breaks that could be seen clearly by laying a ruler along
the ball trajectories shown in the two-page foldout. But the edi-
tors of *Life*, looking at the two-inch sagitta, smaller than the
diameter of the baseball, concluded that the balls did not curve.

Because the sagitta is small, it is difficult to throw a ball with a
diameter of 2.9 inches through three aligned rods such that the
ball will pass to the left of one rod, to the right of a second, and to
the left of the third. However, in the course of arguments in 1870
as to whether a curveball really curves, New Haven native Freddy
Goldsmith performed that feat.

But Goldsmith probably threw a slower curve. For slow curves
thrown with a definite spin rate, the deflection is approximately
proportional to the square of the time the ball is in the air.
Hence, a ball thrown with an initial velocity of 65 mph with a
1600-rpm spin, which takes about 8 percent longer to reach the
plate than the 70-mph pitch, will curve nearly 15 percent more,
even though the Magnus force is actually smaller.

Conversely, according to the model of the Magnus force I

have adopted, it is almost impossible to throw a fastball that curves strongly. As shown in Figure 2.2, the transverse Magnus force that induces the curved trajectory does not increase with velocity for speeds above 70 mph. Moreover, since the faster ball reaches the plate sooner, the force has a shorter time in which to act; for the same transverse force, the ball that travels 10 percent faster will curve 20 percent less.

The slider—sometimes called a "nickel curve" seventy years ago—is a kind of fast curve. Thrown at a higher velocity than the standard curveball, the slider will break less than the curveball and the spin axis is such that the deflection is more nearly left-right than that of the curve—which, at its best, is closer to a pure drop (which is why it is sometimes called a 12-to-6 curveball).

If the slider is half curve, half fastball, then the cut fastball—thrown hard with backspin and some side spin—is half fastball and half slider.

The screwball thrown by Carl Hubbell and others—called the fadeaway by Christy Mathewson and the "scroogie" more recently when Fernando Valenzuela threw it so well—is a kind of reverse curve, thrown by a right-handed pitcher like Mathewson to break away from a left-handed batter a little like a left-handed curveball. Since about 70 percent of batters swing from the right side and eat up a left-handed pitcher's curveball, the screwball, which breaks away from a right-handed batter, is especially useful for southpaws like Hubbell and Valenzuela.

Figure 4.2 shows typical spin directions for different pitches.* Pitches thrown in baseball generally rotate about axes that are perpendicular to the direction of ball's motion—as suggested by the figure—even as such rotation optimally changes the trajectory of the ball in the air. Conversely, cricket bowlers, who usually "bowl" the ball so that it reaches the batter on a bounce, will

---

*It is interesting to observe the curvature in the flight of a table-tennis ball produced by throwing the ball such that it spins in the directions shown in Figure 4.2. Since it weighs ⅟₆₀ as much as a baseball, very small Magnus forces will generate substantial curves in the trajectories of the light ball over a span of 20 feet.

Ball thrown by a right-handed pitcher toward you

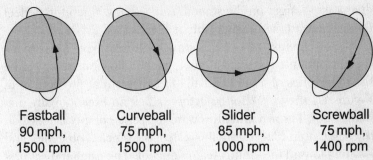

Fastball	Curveball	Slider	Screwball
90 mph,	75 mph,	85 mph,	75 mph,
1500 rpm	1500 rpm	1000 rpm	1400 rpm

FIGURE 4.2: *Ball rotation directions, as seen by the batter, for pitches thrown almost straight overhand by a right-handed pitcher. The arrow shows the direction of rotation—placed so that it also points in the direction of the Magnus force. In all cases, the axes of rotation are in the plane of the paper. The velocities and spins that are listed are only meant to be representative.*

often throw in a manner so as to generate rotation about an axis that is along the direction of motion—like a pass thrown in American football—thus causing the ball to bounce in or away from the batter, as goes the spin.

THE KNUCKLEBALL

The thrown ball can also be deflected as a consequence of the turbulence induced by the stitching on the flow of air past the ball. If the ball is thrown with very little rotation, asymmetric stitch configurations can be generated that lead to large imbalances of forces and extraordinary excursions in trajectory. Low-resistance turbulent flow can be induced by stitches on one side of the ball, while the air will flow smoothly—and with greater resistance—past a smooth face on the other side. I can be more specific. Noting that the drag on a ball is proportional to the drag coefficient, we see from Figure 2.1 that the drag on a smooth ball thrown at a velocity of 65 mph is far greater than the drag on a rough ball. We might then expect that a ball that is smooth on one side and rough on the other will encounter an asymmetric drag force—larger on the smooth side—that will tend to deflect the ball toward the rougher, stitched side.

Measurements* of these asymmetric forces have been made on balls mounted in wind tunnels which allow one to calculate the trajectories of a pitched ball. The diagram at the bottom of Figure 4.1 shows such a trajectory (albeit an exceptionally dramatic one). Thrown initially toward the center of the plate, when the ball is only 20 feet from the plate it is 11 inches off-center and heading toward the visiting team's dugout. The catcher dives desperately to his right to block a wild pitch, and the right-handed batter relaxes, knowing the ball will be a couple of feet outside. Then, confounding the catcher and the batter—not to mention the pitcher!—the ball ducks over the center of the plate for a called strike—and a passed ball.

Since the batter must make his decision to swing from the visual information that he has 300 milliseconds before the ball crosses the plate, it is simply not possible to purposefully hit the breaking knuckleball. He can only swing and hope that he is lucky—or that the pitcher is unlucky and the ball doesn't break so that he connects. Or that if he can't hit the ball, he can perhaps get to first base on a third-strike passed ball.

Indeed, catching the knuckler is an art of its own. Usually the catcher works to block the pitch with an oversize glove or his body as the ball breaks sharply. One catcher, known for his success with knuckleball pitchers, was asked how he managed to handle the pitches. He replied that it was easy, explaining that he just waited until the ball stopped rolling and picked it up.

There were other examples of a fluttering ball—which curves in both directions!—generated by the wind tunnel simulation. In general, the knuckleball pitcher tries to throw the ball with a small rotation (about half a rotation from pitcher to batter) so that the stitch configurations—and the forces on the ball—change on

---

*These measurements were described by R. G. Watts and E. Sawyer in the *American Journal of Physics* in 1975.

the way to the plate. Such a slowly rotating ball was thrown off the knuckles, with the ball held between the forefinger and little finger, by Eddie Cicotte, the great Chicago "Black Sox" pitcher. But most knuckleball pitchers throw the ball off the fingertips or fingernails—which they carefully cut square.

The disadvantage of the knuckleball, from the point of view of the pitcher, catcher, and manager, is that the forces can vary strongly with very small differences in orientation of the ball; hence the pitcher, however skilled, finds it very difficult to control the pitch. If it breaks sharply, it is difficult to catch and leads to too many passed balls. If it doesn't break, it is no more than a batting practice pitch, which is even worse.

THE HOP OF THE FASTBALL

The fastball may be the key baseball pitch; surely it is the most dramatic. How fast can a pitcher throw a ball? It seems that the fastest pitchers can throw the ball so that it crosses the plate with a velocity that approaches 100 mph. In 1946, Bob Feller threw a ball that was measured to have a velocity of 98.6 mph as it passed through a measuring trap about sixty feet away; the ball must have been traveling about 107 mph when he released it. In 1914, a pitch of Walter Johnson's was found to have a velocity of 99.7 mph. Nolan Ryan threw a ball clocked at 100.7 mph. Among the old-timers, Lefty Grove and Smoky Joe Wood* probably threw with comparable velocities, according to contemporary judgment, and Satchel Paige and Slim Jones threw bullets in the Negro Leagues of the 1930s. Grove—and some batters—thought that he was faster than Feller, and Walter Johnson said once, "No one ever threw harder than Smoky Joe Wood." Surely men have thrown hard in the past century of baseball, but I suspect that none threw harder than Randy Johnson does as we enter a new century.

---

*Dr. Joe Wood after receiving an honorary degree from Yale in the 1970s.

Though adequate velocity measurements are possible with instruments as simple as the ballistic pendulum used in 1914 or as sophisticated as the radar guns used today, considerable care is required to get an accurate result with either technique.

Since the ball slows down considerably on the way from the pitcher to the plate, the "muzzle velocity" of the ball—its speed as it leaves the pitcher's hand—is about 8 mph greater than the ball's speed when it crosses the plate. A ball loses speed at the rate of about 1 mph every 7 feet.

It is well known that the backspin applied to the overhand fastball causes the ball to rise—or "hop"; such a ball will be thrown with a backspin of perhaps 1600 rpm and rotate about ten times on its way from pitcher to plate. Though the hop is not likely to be much greater than 7 inches, this is more than enough to trouble the batter swinging a bat two and a half inches in diameter; he must initiate his swing when the ball is less than a third of the way to the plate and the deviation from the hop is only about ¾ of an inch. Like its cousin the curveball, the hopping fastball follows a smooth arc on its way to the plate, but half of the hop deviation occurs in the last fifteen feet of that flight.

Such a hopping fastball, thrown with a lot of backspin, does not "rise" in the sense that it increases its height above the ground as it passes the batter, but it does rise with respect to the trajectory it would have without the spin. The diagrams of Figure 4.3 show the trajectories of two fastballs thrown with the same velocity and original trajectory but with different backspin rates. From the upper diagram, we see that the two trajectories are not very different. But the difference is important. When I plot that *difference* between the heights of the two pitches as they pass the plate, we see that the ball with the greater spin rises above the trajectory of the other ball. And although the difference between the two trajectories is 2½ inches at the plate—about the diameter of the barrel of the bat—that difference is only about ⅝ inches when the balls are halfway to the plate. Moreover, the batter has to make his decision to swing using the information he collects 300 milliseconds before the ball crosses the plate, when the dif-

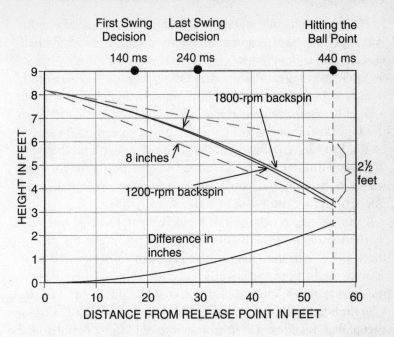

FIGURE 4.3: *The up-down trajectories of fastballs thrown overhand with different backspin rotation rates. Decisions to swing at the pitch are based on information taken from the times noted, as discussed in Chapter 3.*

ference between the two trajectories is hardly more than a quarter of an inch. If the batter bases his swing on the trajectory of the ball with the lesser spin but the pitcher has put extra spin on the ball, he will complain that the ball "hopped" right over his bat—and I would agree.

It might seem surprising that the total rise due to the backspin of the hopping fastball is only about 7 inches. However, that "upward curve" of 7 inches is not very different from the "tailing off" of about a third of the plate width seen in fastballs thrown sidearm (or from a three-quarters-overhand pitch), where the axis of rotation is tilted so the hop has a left-right component that is more clearly seen. Again, the small break follows from the short time in which the force has to act and from the smaller Magnus coefficient at higher velocities, as shown in Figure 2.2.

Some pitchers accentuate the sideways break, releasing the ball with an extra left-right component of spin, and call the half-fastball, half-slider pitch a "cut" fastball.

There are long-standing arguments concerning the magnitude of the rise of an overhand fastball. Can such a pitch, thrown with a lot of backspin, actually rise as it heads toward the plate? From the trajectories shown in Figure 4.3, the answer would seem to be no. But I have pointed out that I do not understand the flight of the ball through the air as well as I would like. Could my arguments be wrong, and could the players who swear that some pitchers throw a rising ball be right?

I can cast some light on the problem by noting that since the 90-mph fastball must fall almost 3 feet on its way to the plate if there is no hop, the effect of the backspin must be such as to "curve" the ball upward about 3 feet if the ball is to rise.* But the magnitude of that hop will be duplicated by a pitcher who throws a 90-mph fastball sidearm, with a sidespin instead of a backspin, except that the curve will be from left to right (for a right-handed pitcher). But who has seen a fastball curve 3 feet? In fact, a hard fastball thrown sidearm (or with a three-quarter motion) seems to curve little more than 6 inches, if that much. Therefore, over-hand fastballs certainly do not rise, and our calculations are not likely to be wildly wrong.

Good softball pitchers, throwing underhand upward to a batter forty-six feet away from a pitcher's plate that is not raised on a ten-inch mound, *do* throw balls that (barely) rise as they pass the batter's armpits. However, the underhand baseball pitchers such as Carl Mays in the 1920s and Dan Quisenberry and Kent Tekulve in the 1980s, who had to throw nearly sixty feet to the batter from an elevation of ten inches, couldn't get the ball to actually rise as it passes the batter.

If the fastball falls about 2½ feet on its way from pitcher to plate, how can thoughtful, intelligent players possibly believe

---

*Indeed, since over-the-top pitchers must throw the ball slightly down to deliver a strike at the belt, the ball must curve up as much as 6 feet if it is to actually rise as it crosses the plate in the strike zone.

that it might rise? Here we have a matter of perception. If one draws a straight line from the hand of the pitcher to the position of the ball as it crosses the plate 56 feet away, as shown by the broken line in Figure 4.3, the peak of the actual trajectory of the ball lies only about 8 inches above that line; the path then differs only about 8 inches from a straight line. The 2½-foot drop is from the line projected from the flight of the ball as it leaves the pitcher's hand. (This is the line that would describe the flight of the ball if the game were played in outer space, where there is no gravity or air.) The player who sees the ordinary fastball as traveling almost in a straight line is not wrong. Therefore we can see that he might reasonably consider that a ball arriving 7 inches above the expected line must have risen.

Part of the reason for misunderstandings of the "rising fastball" among players stems from the foreshortened view of the ball trajectory from the pitcher's mound, the catcher's box, and the batter's box. If the player watches the fastball from the on-deck circle, he will see it fall a lot (about 1 foot in 8) as it passes the batter waist-high. The falling trajectory of the fastball can also be seen occasionally on TV, when the cameraman chooses to show the ball crossing the plate as seen by a camera at right angles to the line from pitcher to plate.

Similar misunderstandings occur over the flight of a throw from third to first by a third baseman with a strong arm. Some players state emphatically that a hard-thrown ball travels in a straight line, but a physicist concludes that the ball must drop about 10 feet. Again, the 10 feet is the drop below the initial line of projection. The peak trajectory of the ball is only about 2½ feet above a straight line drawn from the third baseman's hand to the first baseman's mitt. And this 2½-foot deviation over the 127-foot distance between third and first is surely not far from a straight line. Moreover, since fields are landscaped with the center of the diamond higher than the base paths (so the infield will drain toward the foul lines) and the pitcher's mound is ten inches high, a ball thrown and caught at a height of six feet might pass near the ear of the six-foot pitcher. For the same reason, the pitcher

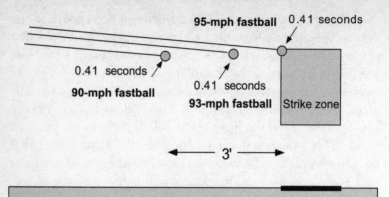

FIGURE 4.4: *The positions and trajectories of fastballs near the plate for balls delivered with the same initial trajectories and spins but slightly different velocities.*

may well have to duck to avoid a catcher's throw to the shoe tops of the second baseman in position to tag a runner attempting to steal second.

Even as differences in the applied backspin generate pitches that have sufficiently different behaviors to confute batters, a small difference in velocity can also trouble them. Figure 4.4 shows the position of a fastball thrown with an initial velocity of 95 mph at the point when the bat must strike the ball if the batter is to get a hit. The positions of balls thrown with slightly lower initial velocities—but with the same initial trajectories—are also shown. We see that the batter who expects a 95-mph fastball will swing much too soon and much too high to hit the 90-mph fastball. Even with but 2 mph taken off the fastball, the batter will hit a foul tip back to the catcher if he swings properly to meet the 95-mph ball. And if the right-handed batter corrects properly for the height, but not the speed, he will hit the 93-mph fastball foul down the third-base line.

The difference in height between balls thrown with different velocities varies with distance, just as the difference between balls thrown with different backspins, as shown in Figure 4.3. Halfway to the plate the difference is about one-fourth as great as at the plate. And, again, it is very difficult for the batter to sense the difference.

The magnitude of the Magnus forces and drag forces shown in Figure 2.2 represent a kind of average over different stitch orientations. Since the pitcher can control those orientations, he can produce somewhat different effects by releasing the ball so that it spins with the axis oriented differently with respect to the stitch pattern. The diagram at the center of Figure 4.5 shows a ball held for the delivery of a "with-the-seams" fastball where two seams pass the equator at every revolution. At the right, the release position for the "cross-seam" fastball is shown, where four seams pass the equator per revolution.

Players generally hold that the four-seam fastball comes over the plate more quickly than the two-seam pitch. While there is no credible objective evidence for a such a difference in ball speed, a small difference is not implausible. As I have noted in Chapter 3, balls delivered with such different spin axes can differ significantly in their aerodynamics. An inspection of the graph of Figure 2.1, showing the variation of the drag coefficients with velocity for balls of different smoothness, shows that the resistance on a 95-mph fastball would be about two-thirds greater if the ball were very smooth. It is certainly plausible that the with-the-seam rotation, showing only two seams per revolution, offers a smoother-than-average profile to the airstream and that the cross-seam rotation might appear rougher than average. If this is the case, we could expect a greater-than-average drag on the with-the-seam (two-seam) fastball and, hence, a somewhat smaller-than-average drag on the cross-seam (four-seam) delivery.

More detailed calculations suggest that the difference—which is wholly in the slowing down of the ball—is quite unlikely to be greater than 2 mph (that is, it would slow down 6 mph rather than 8 mph, with an average difference of 1 mph) as the ball reaches the plate. With these numbers, the four-seam pitch would get to the plate about 7 inches before the two-seam delivery.

There is better evidence suggesting that the four-seam pitch might have a greater hop than the two-seam ball, and that could be important. Again, a 20 percent difference is plausible. This could amount to as much as a 1½-inch difference in pitches that were otherwise identical. Moreover, since the difference varies quadratically with distance, very much as shown in Figure 4.3 for pitches with the same velocity but different spins, it is hard for the batter to pick up.

Some pitchers purposefully throw a "sinking" two-seam fastball that may be less effective in generating a Magnus force. Such a pitcher might hold the ball rather as in the center drawing of Figure 4.5, but with his fingers set on the smooth leather rather than on the stitches so that the ball will slide from the hand with less spin. Such a ball, delivered with less backspin, will then undergo less upward Magnus force and come to the plate lower than the hopping fastball delivered with the same initial trajectory, by as much as 3 inches. With the delivery at his knees, the batter will tend to hit the top of the ball and roll out weakly to an infielder.

The ordinary overhand fastball is thrown such that the ball rolls off the forefinger and middle finger of the pitching hand—indeed, these fingers usually rest on the stitches to accentuate the grip and the spin. If the pitcher places the ball such that he holds its axis between his two widely split fingers at about the first joint, with the same motion he will propel the ball more slowly and with very little spin.* Usually the split-finger fastball is not projected quite as efficiently as the regular backspinning fastball and does not achieve the same velocity. With the same initial trajectory as the hopping 90-mph fastball directed across the letters, the 80-mph split-finger fastball reaches the plate about 16 inches lower and at the knees—and about 6 feet behind the fastball.[†]

---

*If the ball is jammed farther toward the hand, the velocity of the pitched ball is reduced considerably and the pitch is more of a pure change of pace. Such a pitch is often called a "forkball." However, in practice, this nomenclature is used variously.

[†]If they were released by two pitchers at the same time, the split-finger fastball would lag behind the fastball by 6 feet as they crossed the plate.

PITCHING WITH AN ILLEGALLY MODIFIED BALL

From baseball's beginnings, pitchers have enhanced their skills at throwing the ball with different velocities and spins by modifying the ball and the pitching hand. Saliva, sweat from the forehead, hair oil, paraffin, or petroleum jelly surreptitiously applied to the pitching hand aid in the delivery of the slowly rotating "spitball" by allowing the ball to slip from the hand without spin. The main effect of the lubricant is that it allows less skilled pitchers to emulate what a skilled pitcher can accomplish legally—though the spitball can be thrown harder than the knuckleball. The application of a lubricant to the forefinger and middle finger upon delivery of an overhand fastball can produce some of the same effects as the split-finger fastball: The ball will slide off the fingers, accumulating less backspin, and then drop more than the usual fastball thrown with backspin.

Scarring or scuffing the ball can produce asymmetric forces on the ball and aberrant trajectories. Since it is probably impractical to scuff or scrape the surface of the ball in a manner such that the imperfection has as large an effect on the pitch as the ball's stitching does, it might seem that such modifications cannot be very important, and that is largely the case for casual throws. But the highly skilled pitcher* can throw the ball so that the effect of the stitches is symmetric. Then if the ball is scuffed on one end of the rotation axis and not the other, unbalanced forces can be realized that act only in one direction, as suggested in the left-hand diagram of Figure 4.5. The scuffed left-hand side of a properly thrown ball could induce low-resistance turbulence in the air passing by, while the air will pass the smooth right-hand side in a high-resistance smooth flow and the ball will veer to the left—*toward* the scuffed area! Since a deviation of a fraction of an

---

*This scuff-ball effect can also be simulated by a table-tennis ball thrown indoors. To conduct this test, take a square of Scotch tape—perhaps ⅛ inches × ⅛ inches—and stick it on the ball; usually the tape will wrinkle. Then throw the ball with pure backspin as suggested by the left-hand diagram of Figure 4.5 with the tape on one or the other pole of the axis of rotation. The ball will veer, counterintuitively, *toward* the tape.[a]

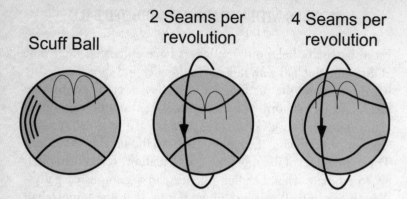

FIGURE 4.5: *The position of the fingers at release time is shown for a typical four-seam overhand fastball at the right, a two-seam release is shown in the center, and a scuff-ball release is at the left.*

inch can change a home run into a pop fly or a double-play ground ball, the controlled deviations a skilled pitcher can induce by disfiguring the ball can be important.

The curve—called the "swing"—of the bowled cricket ball plays a part in that game analogous to that of the curve in baseball pitching. Throwing on the run, the cricket bowler, projecting the ball with an overhand straight-arm bowling motion, sends it to the batter on a bounce—yet almost as fast as a fastball from a baseball pitcher. A cricket ball is about the same size and weight as a baseball, but the raised stitches form a circle about an equator of the ball. Moreover, the bowlers are allowed to work on the cover of the ball, usually shining one side and leaving the other rough. Hence, the bowler does not rely on the Magnus effect but on a kind of scuff-ball effect to project a ball that evades the batsman and knocks down the wicket.

PITCHING IN THE WIND AND ON HIGH

Excepting the recent intrusion of domed stadiums, baseball is an outdoor game where the vagaries of the weather affect the mechanics of pitching. Here we concern ourselves only with the weather's effect on the flight of the pitched ball, leaving the benefits of hot weather on pitching arms to physiologists and psychologists. Moreover, in the contest between pitcher and batter, let us consider for now only the effects of weather on the pitched ball, leaving the problem of windblown fly balls and balls in the sun to outfielders—and managers.

The weather affects the flight of the pitched ball mainly through the velocity of the wind over the diamond. Though there are small effects that follow from variations of temperature and barometric pressure—the fastball is quicker by an inch or two on a hot day and in the lower air pressure before a storm—they will not alter the game significantly. The average wind velocity over the United States—about 10 mph—is surprisingly constant from place to place. (The wind surely blows harder at certain places, such as Mount Washington, but major league baseball games have not been regularly scheduled there.) Moreover, the prevailing wind is from the west and will usually be in the pitcher's face. Of course, the weather is famously perverse everywhere—especially at Candlestick Park in San Francisco, where a balk was called on Stu Miller in the 1961 All-Star Game when the wind blew him off of the mound—and winds can come, betimes, from any direction.

The wind's effect on the pitched ball is subtle but important to pitcher and batter. It is again convenient to consider specifically the effects of a 10-mph wind, though on the playing field of most major league parks the protection of the stands will usually reduce wind velocity somewhat. The effects I discuss all vary nearly linearly with wind velocity; hence a light 5-mph breeze will have half the effect of the standard 10-mph wind.

As an example of the consequences of wind, I consider a standard major league fastball, thrown overhand, that crosses home plate at a velocity of 87 mph 0.42 seconds after leaving the pitcher's hand with a muzzle velocity of 95 mph. I also assume the pitcher applies a typical backspin to the ball of 1600 rpm, which causes the ball to curve up, or "hop," reaching the plate perhaps 6 inches higher than it would have without the spin. With a 10-mph wind in the pitcher's face, the ball will reach the plate slightly later (lagging by about 4½ inches of flight path), traveling slightly slower at about 86 mph, and will arrive at about the same height. With the wind at the pitcher's back, the ball will reach the plate about 4½ inches sooner at a velocity of 88 mph, again with little difference in height.

These effects are all small compared with the precision of pitchers and batters, and not very important. Neither headwinds nor tailwinds affect pitching much. But a crosswind can be a little less innocuous. If the wind is blowing across the field at a velocity of 10 mph, the fastball will be blown about 3½ inches—the diameter of the ball—off course at the plate, enough to trouble the nibblers who are trying to hit corners if the wind is erratic, as it often is. Again, the effects are roughly proportional to the wind velocity.

While fastball pitchers can largely ignore moderate breezes, curveball and junk pitchers will have a lot more trouble. Let us consider specifically a left-right deflection of a wide-breaking curve thrown such that it crosses the plate with a velocity of 61 mph in calm weather 0.6 seconds after leaving the pitcher's hand with a velocity of 70 mph. This particular ball, thrown with a sidespin of 1800 rpm, breaks about 14 inches. Thrown initially toward the inside of the plate, the ball will break over the outside corner. (Actually, the major league pitcher is usually more interested in the vertical, downward deflection than in the transverse break.) With the wind directly behind the hurler, or in his face, there is little significant change. The ball reaches the plate about 9 inches sooner (in flight distance) with the wind behind him and about 9 inches later with the wind in the pitcher's face, but that is not important. The ball also breaks about the same amount, though here our lack of knowledge of balls near this speed leaves some uncertainty. Indeed, a very slow curve may well break more when it is thrown against the wind.

A crosswind, however, modifies the pitch considerably. If the wind is from the north, from third to first, the right-handed pitcher's curve will break about 19 inches toward first base, but if the wind is from the south, the break will be only about 7 inches. Hence a gusty day with crosswinds can cause serious problems for slow pitches. If the wind kicks up, a pitch thrown accurately toward the corner of the plate can turn into a ball or cross right over the center of the plate, to be deposited in the bleachers by the batter as a souvenir for a fan.

Pitchers may also be hurt by high altitudes. A mile high in Denver, the fastball will take a little less time to reach the plate, and gain about 6 inches, but the hop will be about an inch less than at sea level. Even more important, at that altitude the curve will break about 20 percent less. A curve that will break left-right about 8 inches and drop an extra 8 inches (due to the overspin component) at sea level will break about 1⅝ inches less and drop about 4 inches less in Denver. The ball will break less because the Magnus force is smaller and because it will have a little less time *to* break before crossing the plate. The smaller drop follows both from the reduced Magnus force effect and because the faster ball has less time to fall under gravity. Similarly, the knuckleball will dance perhaps 20 percent less.

While moving the fences back may reduce home runs at Denver, the large outfield that results and the reduced break on the pitches leaves Denver as a pitcher's purgatory, if not quite hell.

THE ENERGETICS OF A THROWN BALL

A player's actions in throwing a ball hard are complex, and the analysis of the throw lies more in the realm of physiology than physics, but the calculation of the total energy of the projected ball and an estimate of the rate of energy transferred to the ball (the power) during the throw illuminates the character of the process.

In this estimate, I consider a pitcher throwing a major league fastball such that the ball leaves the pitcher's hand with an initial velocity of 98 mph to cross the plate about 0.4 seconds later with a velocity of 90 mph. I estimate the distance between the point where the ball begins its motion toward the plate and the point of release as 8 feet; that is, the ball is held about 4 feet behind the pitcher's rubber as the pitcher begins bringing the ball forward toward his point of release, which is about 4 feet in front of the rubber. The released ball then has an energy of about $\frac{1}{16}$ horsepower-seconds. Making the crude—but useful—approximation that the force applied by the pitcher to the ball is constant, we find that the throw takes about 0.11 seconds* and the average force on the ball of about 12 pounds generates a mean acceleration of the ball of about 40 times the acceleration of gravity (40 g). With these numbers, we find that the average power transmitted by the pitcher to the ball in the course of the pitch is about 1.5 horsepower! Since the pitcher's body is also put into motion by the contraction of the pitcher's muscles (indeed, the hand and wrist are moving nearly as fast as the ball when it is released), we conclude that the pitcher's musculature must have generated energy at a rate exceeding 3 horsepower during the action of throwing. Such power can only be generated by the large muscles of the thighs and thorax. As pitchers well know, it is important to develop and retain leg and body strength to throw the fastball.

Although the flow of energy from the pitcher's body to the ball is surely complex, and the description of that flow is not within the scope of this work, some comments about a particular aspect of the energy transfer may be in order. In the course of the throw, the elbow leads the hand and ball at a midpoint of the action. At this time energy is stored in the stretching of the tendons of the arm. That energy is then transferred to the ball in the last portion of the throw as the spring-energy of the stretched tendons is released. When age or injury reduce the elasticity of the tendons, the arm goes "dead."

Tall pitchers have a longer pitching trajectory than shorter

---

*This is not the "delivery" time. The time between "commitment"—after which a balk would be called if the pitch were interrupted—and release is more like 0.8 seconds.

(short-arm) pitchers and then apply a little less force on the ball—albeit over a longer distance—to gain the same muzzle velocity. By and large, they probably rely a little less on tendon stretch than short-arm pitchers do and may therefore be a little more durable.

In general, the elasticity of all tissues falls off with age, and the blazing fastball of the young player must be replaced by the craft of the older pitcher. Late in his long and remarkable career,* Warren Spahn, commenting on his life in baseball, said, "When I was young and threw the fastball . . ." At 45, when he retired from the sport, Spahn was probably about as strong as he was at 25 but his arm was not as elastic.

According to the 2001 *Official Baseball Rules* (page 4), "The pitcher's plate shall be 10 inches above the level of home plate. The degree of slope from a point 6 inches in front of the pitcher's plate to a point 6 feet toward home plate shall be 1 inch to 1 foot." The height of the pitcher's plate has been changed at times. If the slope in front is retained at 1 inch to 1 foot, the height is almost irrelevant to the pitcher-batter competition. A decrease (or increase) in height of 2 inches is then precisely equivalent to an increase (or decrease) in the height of the plate—and the strike zone—by 2 inches, and the effect is that of making the pitcher throw each pitch 2 inches higher (or lower).

However, if the slope is changed and the pitcher's front foot lands at a different height, the leverage of the pitcher will change, and *that* can affect the delivery.

PITCHING TACTICS: HIGH AND TIGHT, LOW AND AWAY

In the contest between pitcher and batter, the batter tends to undercorrect. In order to hit the variety of pitches that come his way, he must be poised to hit a medium-fast pitch over the center of the plate—and then adjust to the ball the pitcher actually

---

*Spahn won over 20 games in a season 13 times, the last in a 23-and-7 season in 1963 when he was 42.

throws. But, in practice, against a good pitch, the batter tends not to adjust quite enough. Hence, batters often hit *under* high pitches, sending fly balls to the outfield; tend to hit *over* low pitches, bouncing them through the infield; swing *late* on fast pitches, driving them to the opposite field; and swing *early* on slow pitches, pulling them down the near foul line. They also hit *over* curveballs (which drop as much as they curve) and *under* the hopping, rising fastball.

With these tendencies in mind, the pitcher finds that his best pitches are those where two or more tendencies add and reinforce each other; the worst are those for which they compensate.

I list the normal undercorrection errors in Table 4.1.

When a pitch is such as to lead to compensating errors, the batter hits it squarely. Hence the pitcher's fear of the "hanging curveball," which is simply a curve thrown high in the strike zone. The batter tends to swing under the pitch because it is high, but over it, being a curveball, because it is dropping; the over-under errors cancel, so the ball is hit squarely—and into the bleachers. Pitchers who throw curves—and also sliders—like to get them low in the strike zone.

Pitch	Correction Error	2nd Error
High	Swings under	
Low	Swings over	
Inside corner	Swings late	
Outside corner	Swings early	
Curve	Swings over	Swings early
Slider	Swings over	Swings early
Hopping fastball	Swings under	Swings late
Sinking fastball	Swings over	Swings late
Change-up	Swings early	Swings over

TABLE 4.1: *Pitch types and corresponding likely batter errors*

The fastball pitcher likes to throw "high and tight." The batter has to get the bat around a little early to hit an inside pitch, but it is hard to do that against a fastball. In addition, the batter tends to swing under high pitches and under fastballs. Thus two sets of errors add. Conversely, the change-up—such as the forkball, which is a kind of change—works best thrown low and away.

Of course, a pitcher must mix things up to some extent. If all fastballs are high and tight and all changes are low and away, the batter can select the appropriate response pattern easily. Hence most pitchers work to mislead the batter by mixing up placement and velocity so that the patterns that the batter selects from his mental armory are imperfect—and all with enough pace that the batter doesn't have enough time to perfect that selection.

Arguably, no pitcher had a better record in the decade of the 1990s than Greg Maddux, who threw neither an exceptional fastball nor an exceptional curve. Maddux was reported to have said that he worked to throw each of his many different pitches so that they looked exactly alike as he delivered them and for the first 15 feet of their flight. Hence, when the batter went to select identification and action patterns, he had little information to go on and picked the wrong one often enough to contribute to Maddux's remarkable won-lost record.

Some junk pitchers (Vic Raschi of the 1950s Yankees comes to mind) seem to throw every pitch differently—sidearm, overhand, big curve, small curve, slow, medium, fast, in, out—so that the batter has to review so many patterns that he doesn't often choose the right one.

So batting is difficult, and there are many ways for a pitcher to make it as difficult as possible.

LEFT-HANDED AND RIGHT-HANDED PLAYERS

Managers, players, and fans know that a right-handed batter has more trouble hitting right-handed pitchers than he does lefties, and that left-handed batters have more trouble with left-handed pitchers. Late in the game, right-handed pinch hitters are chosen to face southpaw pitchers, and vice versa. Is this left-right effect real? If so, why, and how big is it?

In Table 4.2, I summarize the results of four years of batting in the major leagues (1984–1987) for all batters are who had over 250 at-bats (AB). The number of hits and the number of at-bats are given for left-handed and right-handed pitchers, while the batters are classified as left (-handed), right (-handed), or switch-hitters.

The switch-hitters hit .2690 ± .0042 against left-handed pitching and .2705 ± .0028 against right-handed pitchers,* for an overall batting average of .2702. The left-handed batters hit the left-handed pitchers (LL) for a batting average of .2568 ± .0036, while against right-handed pitchers (LR) they hit .2855 ± .0023. Their overall batting average was .2778 ± .00194 and their LR-LL differential was .0287 ± .0043. Right-handed batters hit left-handed pitchers (RL) for a batting average of .2813 ± .0052. Against right-handed pitchers (RR), they hit .2649 ± .00168, for an overall average of .2702 ± .00140. The RL-RR differential was .0164 ± .0030.

For the average batter and average pitcher, changing pitchers (or batters) to take advantage of the left-right match is worth .0287 for left-handed hitters and .0164 for right-handed batters. For the average of all batters, the difference between the matched and mismatched hitter-pitcher combinations is .020 ± .0025; the average mismatched batter (RL or LR) has a 7 percent better chance of getting a hit than a matched (LL or RR) batter. So, don't replace a batter who matches the pitcher on left-right considerations with a batter who mismatches but bats .020 points less. And don't replace a good mismatched pitcher with a poorer pitcher who matches. But we all know that. Or do we? And, of

---

*The quoted errors are estimates of "standard deviations." If only random chance variations affected the results, the "real" results would be expected to fall within the stated error about two-thirds of the time. The probability of being off by more than twice the error is about one in twenty.

Bat	Pitch					
	Left			Right		
	AB	Hits	Avg.	AB	Hits	Avg.
Left	25,121	6,451	.2568	69,188	19,756	.2855
Right	56,821	15,989	.2813	119,190	31,571	.2649
Switch	19,772	5,319	.2690	43,235	11,694	.2705
All	101,714	27,759	.2729	231,613	63,021	.2721

TABLE 4.2: *Batting statistics for 1984–1987*

course, there are batters whose left-right difference is much greater—or much less—than the average.

But why do lefties hit righties and righties hit lefties? It's not obvious. Of course, the left-handed and right-handed pitchers throw from different angles, the lefty from the first-base side of the pitcher's box and the righty from the third-base side. But the batter should have no trouble shifting his stance slightly—seldom more than 5 degrees—to compensate for that shift in angles.

Many, if not most, batters attribute the small but significant hitting difference solely to the different break of the curveball. I have had players tell me that they do not care whether the overhand fastball is thrown left or right but that they find it more difficult to hit the curve that breaks out than the curve that breaks in. And indeed, right-handed screwball pitchers tend to throw screwballs to left-handed batters and curves to right-handed batters so that all batters, swinging from the left or the right, see balls that break away from them.

But why do batters hit balls curving in better than those curving out? The out-curve that breaks over the plate is thrown at the batter. Some players say that such a curve freezes the batter briefly, thus reducing his adjustment time, since he now has one more decision to make—to hit the deck or not. But there are other factors that might play a role.

I have noted that in the contest between pitcher and batter, the batter tends to undercorrect. The compensation errors of Table 4.1 may be important in considering hitting in-curves and out-curves. Batting against the curveball, the batter tends to swing too quickly at the relatively slow pitch and tends to underestimate the in-out curve deviation. These errors add up for out-curves but counteract one another's effects for in-curves. Hence, the in-curves may be a little easier to hit, accounting for the small advantage batters have when they face a pitcher throwing from the opposite side. Figure 4.6 illustrates this effect.

For a tailing fastball thrown by a right-handed pitcher, breaking a little away from the batter, the right-handed batter's error in alignment and his late swing tend to add up in a manner similar to that for the curveball, while the errors compensate for a left-handed batter. Again, lefty batter hits righty pitcher and righty batter hits lefty pitcher best.

Left-handed hitters hit a little better, .0072 ± .0024, than right-handed hitters. Nominally, left-handed batters have two advantages: 1) they are a step nearer first base, and 2) they face

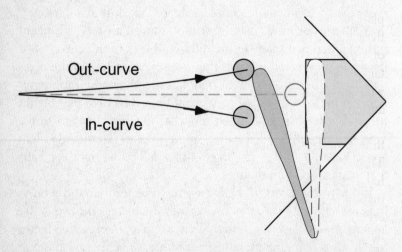

FIGURE 4.6: *The batter aims to hit a fastball (broken ball and bat) but the curveball arrives later with a deflection (solid lines). The right-handed batter swings too soon at the slower pitch and hits the in-curve solidly, near the sweet spot, but misses the out-curve or hits it weakly, near the end of the bat.*

more opposite-handed (right-handed) pitchers, who are presumably easier to hit than same-handed (left-handed) pitchers. Hence, we understand their higher batting averages. But do we?

Practically all who throw left-handed bat lefty (Presidents of the United States, such as George H. Bush, excepted). Though many right-throwers bat left, about 70 percent bat right-handed, which suggests that there is some coordination or strength advantage in batting as you throw. The skill positions—second base, shortstop, third base, and catcher—where excellent fielding can compensate for mediocre batting, are reserved for right-throwers who tend to bat right. Southpaws *have* to hit well to play baseball field positions. Hence part of the L-over-R batting advantage must derive from the different positions available to left- and right-throwers.

An educated guess suggests that *all* of the left-right batting difference is due to this position difference. Hence, since their advantages are real, left-handed hitters in baseball need be a little less naturally gifted than their right-handed counterparts.

Switch-hitters bat about equally well from either side of the plate. From a casual inspection of the records, it seems that more than 90 percent of switch-hitters throw right. Hence, the equality of their left-right batting averages suggests that the natural advantages—strength? coordination?—that accrue to right-handers who bat as they throw just compensates for the nearer-to-first-base advantage of batting left-handed.

But there is one more factor: About 30 percent of the pitches thrown during the four years studied were thrown by southpaws. The near equivalence of total batting averages against left-handed (.2729) and right-handed (.2721) pitchers suggests that managers and general managers do a good job of selecting pitchers without left-right prejudice. However, the proportion of left-handers in the male population is not likely to exceed 15 percent. Hence, left-handed pitchers are probably intrinsically slightly inferior to right-handers but are, as a group, slightly favored by the asymmetric configuration of the game and the asymmetric distribution of left-right batting. (Of course, this leaves open the

possibility that left-handers are fundamentally superior to righties. And as a southpaw myself, that's my position.)

Those of us who follow baseball with inordinate attention are
called "fans," presumably from "fanatic." We answer that to
understand the game is to understand life. That illumination is
seldom direct, but the remarkably complete and voluminous
data on all men—approximately 15,000 of them—who ever
played as much as one game of major league baseball may
address, uniquely, a specific question. The data, which includes
birth and death dates and whether they threw from the left or
right, allows an analysis of left-right mortality that may have no
parallel. This is especially interesting inasmuch as some have
concluded—on the basis of suspect data—that left-handed men
do not live as long as right-handed men.

There is a problem in defining what it is to be left-handed.
Those who are born southpaws have to learn to do many things
with their right hand growing up in a right-handed world, and
their natural-handedness is often blurred by social and environ-
mental pressures. So how does one define a left-hander? Better
not by the hand used to write or eat. I suggest that the hand used
to throw a stone or ball is the most nearly culture-resistant index.
Learned early, forgotten never, the left-hander always throws
with his left hand. But what files record this handedness? Baseball records do, perhaps uniquely, and for a span of over a hundred years. I took some of that data, from *Total Baseball*,* to see
if left-handed and right-handed ballplayers have different life
expectancies.

Table 4.3 shows the birth and death dates and throwing-
handedness for 479 right-handed and 159 left-handed players
selected randomly from the many thousands listed in *Total Baseball*. The majority of these players were not stars, of course;
indeed, many came up to the majors for only a cup of coffee, but
since we are studying longevity, not baseball, excellence, this is
not wholly relevant. About 13 percent of all field players and

---

Total Baseball, eds. J. Thorne and P. Palmer, Warner Books, New York, 1991.

about 25 percent of all pitchers were southpaws. Limited in numbers as that is, the set is sufficiently large that the statistical uncertainties are matched by systematic effects; that is, a larger sample cannot be expected to give an appreciably more significant result from the limited information we have on the players.

My finding was that *there is no significant difference between the longevities of left-handed and right-handed baseball players.* The left-handed players in the sample actually lived longer by about a year, on the average, than the right-handed players, but that small difference was not outside the statistical and systematic uncertainties in the study. On the other hand, the mean difference of about three years between the lifetimes of pitchers and field players is suggestive of a truly shorter average life for pitchers (some—envious?—field players attribute this to the fact that with their frequent days off, pitchers are able to drink more), but that conclusion would have to be confirmed by a more careful study before acceptance. The mean birth date is important only inasmuch as it is about the same for the two classes of players showing that certain biases were not important. Likewise, the actual values of the mean lifetimes have no simple meaning. Nevertheless, the mean age at death of about sixty-seven for men born at a

	Right-Handed	Left-Handed
Field Players		
Number	243	77
Birth date	1891.25 ± 1.02	1895.1 ± 1.82
Age at death	67.7 ± 0.9	68.95 ± 1.6
Pitchers		
Number	238	82
Birth date	1891.9 ± 1.04	1892.1 ± 1.83
Age at death	64.7 ± 0.9	65.75 ± 1.55

TABLE 4.3: *Mean age at death and birth date for left-handed and right-handed field players and pitchers. The listed errors are standard deviations.*

mean time of 1892 who lived long enough to be about twenty-two and play baseball suggests that baseball players live as long as or longer than the rest of the male population—be those players left-handed or right-handed.

TECHNICAL NOTE

a) The nominal similarity between the action of the table-tennis ball and the scuffed baseball may be a little misleading. Since the Reynolds number for the small ball, traveling at rather low velocities, is only about 10,000 (corresponding to a baseball velocity of only 5 mph), the disruption of the boundary layer, important to the knuckleball and scuffed ball, is probably over-shadowed by the Bernoulli effect for the table-tennis ball. That is, the air flowing over the Scotch tape has a longer path and must move faster, thus generating a lower pressure, which results in a deflection of the ball toward the tape. The similar effect of the speeding up of the air passing over the protuberant stitches of a baseball would not seem to generate quite enough force to account for the deflections of the heavier baseball.

5

BATTING THE BALL

THE BAT-BALL COLLISION

For a well-hit ball, the bat-ball collision takes place in a small "sweet spot" region that is near the *center of percussion* of the bat* and near the nodes of the lowest vibrational amplitudes of the bat.† An impact in this region on a free bat transfers minimal momentum—or force—to the handle and excites a minimum of energy-absorbing bat vibrations. Such a collision with the ball takes place over a time of about 1/1000 of a second, during which the contact point on the bat moves only about an inch and the handle moves a much smaller distance. The lack of response at

---

*The center of percussion (COP) is a point conjugate to the point on the handle where the bat is held. A ball that strikes the stationary bat between the COP and the handle tends to drive the handle in the direction of the ball's motion. A ball that strikes the end of the bat beyond the COP tends to drive the handle forward, opposite to the direction of the ball. A ball that strikes the COP has no effect on the handle.

†The nodes of a vibration of a linear system, such as a taut string or a baseball bat, are the points where the amplitude is zero. For the second harmonic vibration of a taut string tied at the ends, there will be a node at the center of the string and, of course, nodes at the tied ends. A node near the trademark of the fundamental vibration—at about 170 Hz—is the baseball player's sweet spot. A second node of the 170-Hz vibration is near the handle. Conversely, the antinode is the point of maximum amplitude between two nodes. The Herz, or Hz, is the measure of oscillation frequency equal to cycle per sceond.

the handle, and thus the lack of effect on the collision of forces applied to the handle, means that at the moment the bat strikes the ball, the bat can be considered as a free piece of wood, with the clamping effects of the grip of negligible importance. It is this property of the swing of the bat that allows us to divide the action neatly into the complex physiological actions that transfer energy from the body to the bat and the relatively simple processes that transfer energy from the bat to the ball. I address only the bat-and-ball interaction here.

Very large forces are required to change the motion of the 5⅛-ounce ball from a speed of 90 mph toward the plate to a speed of 110 mph toward the center-field bleachers in the 1/1000 of a second of bat-ball contact. Hence, for a long home run, the force on the ball reaches a value near 9000 pounds, with an equal reactive force on the bat. Such forces distort bat and ball: The ball is compressed to about half of its original diameter, the bat is compressed about ⅕₅ as much. The collision is not elastic—much of the energy of ball and bat is dissipated in frictional heat.

Treating the bat as a moving free object with known weight, weight distribution, and velocity, and knowing the weight and velocity of the ball, one can calculate the initial velocity of the batted ball from the laws of mechanics, given the inelasticity of the ball and bat. The ball may be considered as a spring, a spring with friction; the bat applies force to the ball, compressing it, and the ball exerts force on the bat upon regaining its original contours. The recoil from this exerted force propels the ball away from the bat. But the spring is dampened; the recoil force is less than the compressive force as some of the collision energy is absorbed by friction.

Figure 5.1 shows the results of measurements by Paul Kirkpatrick of the force vs. compression distance for a simulated bat-ball collision, together with similar diagrams for the compression of a golf ball and the compression of the bat. (Complete *stress-strain* measurements were not made on bats, and the curve shown for bats is an estimate presented to show graphically the small energy loss created by the compression of the bat.) The

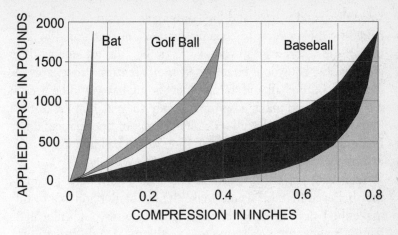

FIGURE 5.1: *The distortion of balls and bats as a function of the applied force for the cycle of compression and expansion. The displacement here is the change in the distance between the impact area and the center of mass of the bats or balls (for a small symmetric distortion, this would be a change in the radius).*

upper curves of each set plot the force vs. distortion distance for compression; the lower curves show the force vs. distortion distance for the following expansion. The area under the upper curve is proportional to the energy absorbed by the ball in the motion; the area under the lower curve is the energy returned by the ball, expanding against the bat in resuming its spherical shape. For an ideal, perfectly elastic spring, the two curves will coincide; the energy released from the spring upon expansion is equal to the energy absorbed upon compression. The area enclosed by the two curves is proportional to the energy dissipated or lost in friction. For the ideal spring this is zero. As shown here, in expansion the baseball returns only about 35 percent of the energy supplied in compression. The golf ball, much closer to a perfect spring, returns about 75 percent of the compressive energy when struck by a driver.

The wooden bat is actually more elastic than the ball,* but

---

*If you drop a ball and a bat, held vertically an equal distance above concrete, the bat will bounce higher than the ball—and an aluminum bat will bounce much higher than a wooden bat.

since the wood is hard and the compression is small, the energy stored by the wooden bat is only about $\frac{1}{50}$ of that stored by the ball. If the wood were just as elastic as the ball, the energy that distorted the bat would be returned as efficiently as the energy of distortion of the ball and the hardness of the bat would be irrelevant. If the wood making up the bat were so weak that the distortion was not returned at all—i.e., if the ball put a permanent dent in the bat—all that energy of distortion would be lost and the ball would not leave the bat with quite as much velocity.

Years ago, players honed their bats with a ham bone because they believed that this would make the bat harder and as a result make the ball fly off it faster. Sometimes batters would (illegally) hammer nails into the hitting area so the ball would strike iron. Since the ash bat stores so little energy upon impact, and returns that energy more efficiently than the energy disposed of in distorting the ball, hardening the bat is pointless. Indeed, if the hitting area were armored with a thin sheet of absolutely hard and rigid material, the well-hit ball would probably go slightly less far. Energy would be lost as the distortion of the more elastic wood would be replaced by an increased distortion of the less elastic ball. But the effect would be small at best. If the infinitely hard armor-plated bat were substituted for the bats now in service, this would decrease the distance of a 400-foot home run by only about 2 feet. The benefit from any effort by players to harden their bats—legally or illegally—must be considered largely psychological.

The inelasticity is usually described in terms of a *coefficient of restitution* (COR), which is the ratio of the velocity of the ball rebounding from the surface of a hard, immovable object to its incident velocity. Consequently, the COR is equal to the square root of the proportion of the collision energy returned to the kinetic energy of the ball's flight.

The Rawlings Company, which makes the balls used in major league baseball, has a sophisticated test facility at its manufacturing plant in Costa Rica. There, with oversight by Major League Baseball, the company regularly tests the COR for samples of the manufactured balls. For baseballs traveling 85 feet per second (58 mph) striking a wall of ash boards backed by concrete, they find a rebound velocity of 48 feet per second and thus a COR of $^{48}\!/_{85}$, or 0.565. They hold that COR closely. An independent analysis made in 1987 found ball-to-ball variations that translated to a difference of only a few feet in a 400-foot home run.

Since the energy of the ball is proportional to the square of the velocity, the rebound energy of the 85-feet-per-second ball is about 0.563×0.563, or 0.32 times the incident energy. Therefore, the collision is quite inelastic; 68 percent of the energy is lost to friction (satisfactorily close to the 65 percent found in the static measurements of Kirkpatrick designed to address the ball-bat collision).

For balls hit by a bat near the sweet spot, the bat takes up very little of the collision energy, so the ball-bat COR is probably near that of a ball hitting a hard wall. There may be small differences, however, since the distortion of a ball hitting a cylindrical bat is a little different from the distortion of its hitting a flat wall.

My colleague R. C. Larsen augmented the results at 58 mph made by Rawlings and measured the COR of a few 1988 National League balls to be 0.588 for an initial ball velocity of 25 mph and 0.584 for 18 mph. A decade later, he found about the same values for 1998 balls. The balls had not changed over the decade.

Isaac Newton concluded through his experiments that the coefficient of restitution is nearly independent of velocity for uniform spheres. However, Briggs found that the coefficient of restitution of a golf ball fell off considerably with increased collision velocity, and the COR for baseballs probably also decreases as the collision velocity increases. Measurements have now been made elsewhere that are relevant to higher velocities. While the COR values for balls hitting a flat wall are not likely to be quite the same as those for balls hitting a cylindrical bat, any differences are probably small. Hence, the wall-ball values of the COR plotted in Figure 5.2, used for the bat-ball collisions considered here, probably describe bat-ball collisions to a satisfactory approximation.

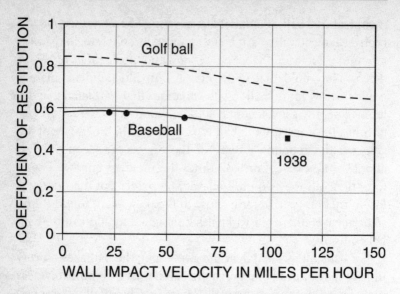

FIGURE 5.2: *Results of measurements of the coefficient of restitution of 1987 and 1997 baseballs striking a rigid flat surface are shown by solid circles, and the value for a 1938 baseball—measured in 1942—is shown as a square. The solid line shows my estimate of the variation of the COR for modern-day baseballs as a function of the impact velocity used in my calculations. The broken curve shows the variation of the COR for a typical golf ball.*

With the insight into the stress-strain characteristics of the baseball provided by the Kirkpatrick measurements shown in Figure 5.1, together with the information concerning the coefficients of restitution shown in Figure 5.2, we can use the simple physical principles of the conservation of energy and momentum to construct a rather reliable description of the ball-bat impact. This applies to the ball that collides with the bat in the sweet spot area and that does not much excite vibrations in the bat. The relevant information is contained almost completely in the model stress-strain diagram of Figure 5.3. Note that the "distortion" in the figure refers to the displacement of the ball's surface relative to its center of mass and is then about equal to half of the compression of the diameter of the ball.

Using the information depicted in the graph of the figure, I

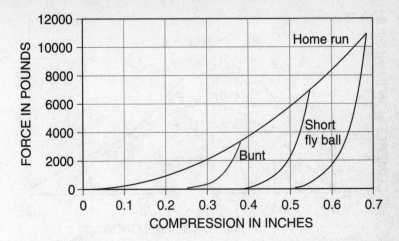

FIGURE 5.3: *Plausible stress-strain cycles for a baseball struck by a bat for impact velocities that correspond to typical bunts, short fly balls, and home runs. The compressions correspond roughly to changes in the radius of the balls.*

derive the variation of force with time and the variation of ball compression with time[a] during the ball-bat collision that are shown in the graphs in Figure 5.4. Note that although the bat-ball contact extends over about 1/1000 of a second—a millisecond— the interval of strong impact is only about half a millisecond.

The diagrams in Figure 5.5 provide a simplified picture of my estimates of the cross section of the ball upon maximum distortion for the three impact conditions considered. A ball's elasticity for the small distortions that occur in bunts—or fungo hitting—is largely determined by the character of its outer yarn windings and is not much affected by the properties of the ball's core. Conversely, the elasticity of a ball that is very distorted by the impact that results in a home run off a fastball is significantly affected by the character of the ball's core.

The strength of the ball-bat impact can be defined usefully in terms of the change in velocity of the ball. The velocity change corresponding to the ball's striking a wall at a velocity 58 mph (85 feet per second) and rebounding at a velocity of 32 mph (48 feet per second) is simply 58+32, or 90, mph. Hence, this is about the

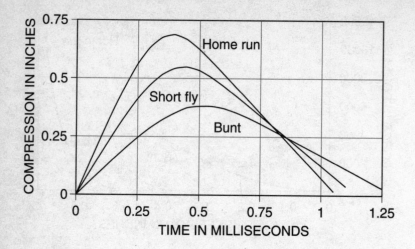

FIGURE 5.4: *The top figure shows the variation in time of ball compression in a ball-bat impact. The lower figure shows the variation in time of the force between the ball and bat. The compressions are distances from the compression face to the center of mass of the ball and thus correspond roughly to changes in the radius of the ball. The bottom figure shows the variation of the bat-ball force as a function of time during the collision.*

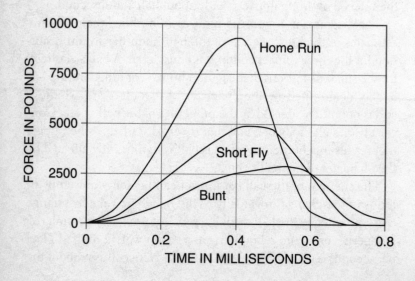

same impact as a baseball experiences when an 80-mph pitch is bunted to go 10 mph—or when a stationary ball is given an initial velocity of 90 mph toward the outfield by a fungo bat. Briggs's measurements of the COR of 1938 baseballs at an impact velocity of 89 mph showed a velocity change of about 130 mph, which is about the same change as that which occurs when a 200-foot Texas leaguer is hit off of a 65-mph curveball. But when a 400-foot home run is hit off an 85-mph fastball, the velocity change of about 195 mph corresponds to the impact of a ball traveling with a velocity of about 135 mph striking a wall.

Major League Baseball tests of the elasticity of balls take place at velocity reversals characteristic of bunts. These tests are certainly a useful check on the consistency of the manufacturing process, but they do not establish the constancy of the ball's elasticity for the long home run.

THE SPIN OF THE BATTED BALL

For almost all bat-ball collisions, the line of momentum transfer through the ball does not coincide with the final direction of the ball's flight, since the bat does not strike the ball perfectly squarely.

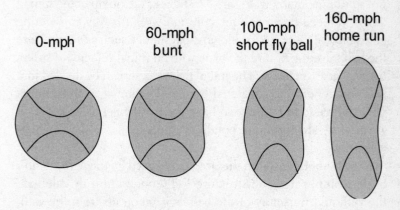

FIGURE 5.5: *Approximate cross sections of balls at the maximum compression point during the bat-ball impact for several impact velocities.*

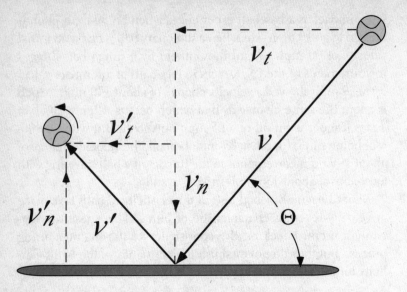

FIGURE 5.6: *The diagram shows initial and rebound velocity components for a ball striking a flat surface at an angle. The curved arrow shows the direction of spin of the rebounding ball.*

The ball is rarely lined straight back to the pitcher, but is usually hit to the left or right, and on the ground or in the air. From such off-line impacts, the ball will leave the bat spinning, though it was not spinning originally. Figure 5.6 suggests the origin of the spin of a ball induced by an oblique collision with a bat by considering collisions with a simple, stationary, flat surface such as a wall. Here the ball traveling toward the plate with an initial velocity, v, strikes the wall at an angle, Θ. The initial ball velocity can be divided into two components, as suggested by the diagram: v_n is the component normal to the surface and v_t is the part tangential to the wall; v'_n and v'_t are the corresponding components of the rebound velocity v'.

We do not understand the stresses induced by the deformation of the complicated ball structure well enough to truly calculate the spin and trajectory of the ball as it rebounds from the wall, but we can deduce a sensible recipe that will allow us to make reasonable estimates of the state of the rebounding ball.

First we can assume with confidence that the normal component of the rebound velocity is simply reduced from the incident normal velocity by a factor equal to the COR. For a value of the COR of 0.5, the normal velocity of the rebound v'_n would be just half of the initial normal velocity v_n. If the ball struck a very slick wall obliquely and skidded off it, the transverse velocity v_t would be unchanged by the collision. However, the forces between the ball and wall are so large for the collisions that interest us that we expect the ball to *roll* on the wall in the course of the collision, because the friction between the ball and the wall acts to stop the surface of the ball where it touches the wall. Then the ball would come off the wall spinning and some of the transverse energy of the ball would go into the spin energy, reducing the transverse velocity of rebound. Moreover, this rolling of the highly deformed ball must take up some additional energy in internal friction. We would expect that this energy loss would be less than, but related to, the dissipative energy loss that reduces the normal component of the rebound velocity. If the value of the COR were near 1 and there was no frictional reduction of the normal rebound velocity, we would expect no reduction of the tangential velocity from frictional losses.

For the normal component of the velocity, which is reversed upon impact, if the COR is 0.5 the rebound velocity will be half of the initial velocity. But the transverse component of the rebound velocity, which is not changed in direction, will retain about three-fourths of the initial transverse velocity.

I calculate the spin of the batted ball struck obliquely by the bat using recipes that follow these considerations.[b] Although these kinematic relations refer ostensibly to the collisions of the ball with a stationary flat surface, within the uncertainties of the recipes they should adequately describe the collisions of the ball with the cylindrical bat when properly translated to the system of the moving bat.

Measurements by Briggs, cited in his National Bureau of Standards report, which show that any spin of the incident ball—i.e., from the pitcher—is largely removed in the course of the bat-ball impact are substantiated by the observations of batters who speak of "straightening out the curveball." Hence, although the overhand fastball, thrown with backspin, will come off the bat with a little less backspin and at an angle about a degree lower than the same ball thrown without spin—and the curveball, thrown with overspin, will come off the bat with a little more backspin and about a degree higher—we do not consider these small effects in our model. (A degree up or down represents a foot up or down as the ball passes the pitcher or two feet as the ball sails past second base.)

In the explicit calculations, I assume implicitly that the bat is swung approximately in a horizontal plane.* Then the ball pulled by a right-handed batter toward the left-field foul line will strike the bat at an angle of about 20 degrees from normal and the ball will spin counterclockwise, as seen from above, as it leaves the bat. If the bat strikes the ball below the ball's center, the fly ball will take off with backspin; if the bat strikes the ball over the center, the ground ball that results will move across the infield with overspin. For a ball hit with a greatly tilted swing— such as the golflike swing at a pitch inside at the knees—the ball spin-axes are changed in an obvious fashion.

Since the spin of the ball affects its flight significantly, it is important to consider the spin induced by batting the ball in spite of the uncertainties of such calculations. Furthermore, just as there are uncertainties in our knowledge of the initial spin of batted balls, there are uncertainties in our knowledge of the effect of the spin on the ball's flight (Chapter 2). Since the spin is important, however, we check the results of the models against that which we know from observations of the game. As in pitching, the effects of up-down spin (usually backspin) are difficult to disentangle from the effects of gravity, and then difficult to observe unequivocably. But the curves from sidespin are better defined. Hence I calculate from our bat-ball recipe, and from my descrip-

---

*Most batters direct the planes of their horizontal swings—at waist-high balls—about 10 degrees upward to match the descent of the average pitched ball as it crosses the plate.

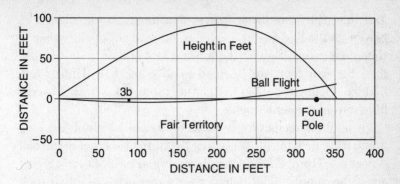

FIGURE 5.7: *The trajectory of a ball hit down the left-field foul line over the third baseman's head, showing the curvature produced by the sidespin induced by the bat.*

tion of the Magnus effect in Chapter 2, the curve of a fly ball hit down the left-field foul line—over the left-field fence—and show the results in Figure 5.7. The ball carries an initial sidespin (clockwise, as seen from below by the third baseman) of about 2400 rpm. I find that the ball, passing over the third baseman a few feet inside the bag, curves about 12 feet foul at the 325-foot foul pole and lands in the stands 385 feet from home plate, now "foul" by nearly 25 feet. Since this behavior is in rough accord with what we have all seen many times, we can conclude that my recipe is probably good within an uncertainty of, perhaps, 30 percent.

In these calculations, I assume that the motion of the face of the bat is at right angles to that face and that all of the sidespin of the ball follows from the angle of incidence of the ball with the bat. Using the analogy of a golf shot, the club face is square with respect to its direction of motion. If this is the case, a ball hit to right or left field will always curve toward the foul line and a ball hit to center will have very little sidespin. However, many batters swing in a manner that the bat is moving along its axis when the ball is struck, giving a sidespin even to balls hit back toward the pitcher, analogous to a slice or a hook in golf. From watching the flight of the ball, it seems that most batters hook a little—a right-handed batter hits the ball with a little extra counterclockwise spin (as seen from above), inciting a curve toward the left-field foul line. But some batters slice. Catcher Rich Gedman, who played for the Red Sox in the 1980s, hit with an "open" stance (a slice stance in golf), with the forward foot farther from the plate than the rear foot, and seemed to slice the ball. Hitting left-handed, Gedman's drives to right field sometimes even curved a little toward the center fielder.

Using these recipes for the spin of the ball, we find that long home runs are typically hit so that the initial backspin of the ball is near 2000 rpm. Ground balls have similar rates of overspin that affect the bounce. The statement is sometimes made that such balls gain speed as they bound through artificial turf infields. At 2000 rpm, the rotational velocity of the ball's surface at the spin equator is about 17 mph. A ball can only gain speed on a bounce, if that rotational surface velocity exceeds the regular linear veloc-ity of the ball, which will be upward of 50 mph for balls hit hard enough to escape the infielders. Hence, ground balls will not speed up even with an overspin as great as 4000 rpm—but they will slow down less than balls with little overspin.

Rarely—but they are remembered in the nightmares of center fielders—line drives are hit to center field such that the ball has almost no spin, darts about like a knuckleball, and is almost uncatchable. Balls hit to left or to right field will generally have enough sidespin to ensure a smooth trajectory.

EFFECTS ON THE DISTANCE BALLS ARE HIT

Obviously, a ball travels farther if it is struck by a bat that is swung faster—or "harder." Less obviously, the fastball hit by a full swing of the bat will travel faster and farther than a slow change-of-pace pitch hit with the same swing. But just how big are these effects?

For the sake of definitiveness, in the following discussions I consider a bat 34 inches long and weighing 32 ounces striking a waist-high pitch such that the plane of the swing lies at 10 degrees from the horizontal (i.e., the bat is swung slightly upward) to drive a ball at an angle of 35 degrees. Among the vari-

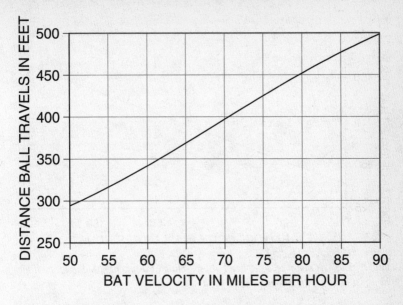

FIGURE 5.8: *The distance an 85-mph pitched ball can be hit by bats swung at different velocities.*

ety of real swings, some vary considerably from this model; the swing at a ball passing over the inside corner of the plate at the knees often seems more like a golf swing. However, the salient points illustrated by the model will be valid for all of these swings.

The graph in Figure 5.8 presents the distance that a ball crossing the plate at a velocity of 85 mph will travel when hit solidly with a full swing as a function of bat speed. I assume further that the ball is hit near the point of maximum hitting efficiency (the sweet spot discussed in detail in Chapter 6), and the quoted velocity is the velocity of that spot—the end of the bat will be traveling about 15 or 20 percent faster. Moreover, I assume that the ball is hit toward center field. Thus, a bat speed of about 70 mph is required to hit a ball 400 feet.

For a given bat speed, a solidly hit fastball goes farther than a well-hit slow curve.* Figure 5.9 shows the distance the ball trav-

---

*Similarly, the fastball hitting a concrete wall will rebound harder than the slower curveball hitting the wall.

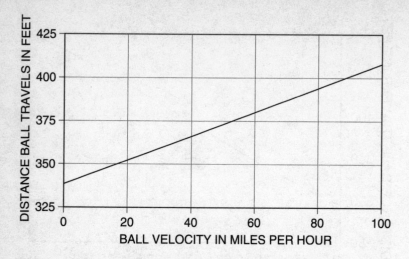

FIGURE 5.9: *The distance a pitched ball, traveling at different speeds, can be hit by a bat swung with a velocity of 70 mph.*

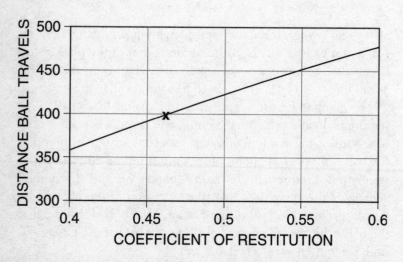

FIGURE 5.10: *The distance a batted ball can travel as a function of the value of the coefficient of restitution of the ball and bat if the bat velocity is 70 mph and the ball velocity is 85 mph. The cross marks an estimate of the value of the COR for modern-day balls and bats.*

els off a bat being swung with a speed of 70 mph, striking balls that cross the plate at different speeds. The swing that hits a fungo (but with the bat described above, not a fungo bat) 340 feet will drive the 90-mph fastball 410 feet!

Of course, a lively ball will travel farther than a dead ball. Figure 5.10 shows the distance a hard-hit home run ball will travel for different values of the COR under conditions such that an 85-mph fastball will be hit 400 feet if the COR has a value of 0.460. Balls can be made that are much more lively and that would therefore travel much farther. A baseball (perhaps wound with rubber like a golf ball) with a typical golf ball COR of 0.85 would travel nearly 600 feet from the same blow that would have sent an ordinary ball 400 feet.

Major league specifications require the COR, measured at 58 mph, to lie between the values of 0.514 and 0.578, an allowable deviation of about ± 5 percent. If that proportional variation holds for the smaller values of the COR that we believe are relevant at greater-impact velocities, we might expect ball-to-ball deviations in flight length of the order of 10 percent. In fact, based on measurements of 72 balls that struck a wall at an impact velocity of 58 mph, the mean ball-to-ball COR variation of balls now in use seems to be no larger than 0.005, corresponding to a deviation in distance of the 400-foot home run of only about 4 feet.

In Chapter 2, I noted that the conditions in which balls were stored could affect the elasticity of these balls and consequently the distance they were likely to be hit—e.g., balls stored under conditions of extreme humidity could be expected to provide long flies as much as 30 feet short of the distance that might be expected for normal balls. The elasticity of balls stored under extremes of cold or heat can be affected also. There are apocryphal stories, going all the way back to John McGraw when the home team supplied the balls to the umpire one by one, of home team managers storing on *dry ice* the balls to be given the umpires when the visiting team was at bat! Those balls were supposedly taken off the dry ice a few hours before game time so that the cover could warm up, leaving umpires' suspicions unaroused— the core of the ball would remain cold and dead! Since McGraw managed, the rules have been changed so that *all* balls—still supplied by the home team—must be given to the umpires two hours before game time; hence, any doctored ball is as likely to be used by the home team as by the visitors, rather evening things up.

R. C. Larsen and I have made some simple measurements[c] which suggest that the temperature effects can be substantial; deep-freezing to −10° F would seem to take about 25 feet off a 375-foot fly ball, reducing many a home run to a loud out on the warning path. Such a ball hit on the ground would also skip through the infield a little more slowly. Conversely, storing the balls in a warming oven at 150° F would seem to inject enough rabbit in the ball to take the 375-foot fly ball over the fence to land about 400 feet from home plate.

In general, it seems that a swing of the bat that would hit a ball 375 feet at a temperature of 70° F would drive the ball an extra 3 feet for every 10° increase in the temperature of the ball. Conversely, the ball would travel about 3 feet less for every 10° decrease in the temperature of the ball. Even on very cold days, the ball temperature probably seldom drops very much below 60°, since the ball in play was probably stored at room temperature before the game and will be kept from cooling too much by the handling of the pitcher. It is more likely that the ball will reach the ambient temperature on a really hot day; if the temperature on the field is 100°, the balls might reach nearly that temperature as well.

Though there are still stories of teams that rely on singles, speed, and defense storing all balls in a deep freeze until a few hours before game time when they play a slower team that depends more upon power and home runs, I am not a believer. Since the careers—and salaries—of the home team batters aren't helped much by the lower batting averages that must result from hitting a cold, dead ball, any cooling of the balls must be kept secret—impractically secret, to my mind.

It is interesting to note that a ball hit toward a foul line will go somewhat farther than a ball hit to dead center, though the bat speed is the same. This follows because the ball must strike the bat at an angle (as shown in Figure 5.6) and the tangential part of the ball velocity, v_t, is not reversed by the bat-ball collision but retained, with little loss in the blow. With the same bat speed, a batter will hit a 90-mph fastball about 403 feet toward the left-center power alley with a swing that would hit the ball 400 feet to dead center field. And if hit along the foul line, the ball would travel about 411 feet.

Table 5.1 lists a number of these effects and estimates of their influence on a standard 400-foot home run. I also show a projection of the effect on the probability of home runs, using my estimate that for the average ballplayer, the probability of his hitting a home run increases about 7 percent for each percent increase in the distance a fly ball travels.

Although backspin adds distance to a long fly ball at a rate of about 1 foot of flight per 100 rpm of spin, there is probably little or nothing that a batter can do to change that spin. Any plausible batting stroke that takes a ball crossing the plate at a small downward angle of about 10 degrees and propels it upward at the angle near 35 degrees required to clear outfield fences will provide the ball with substantial backspin. Although the Ted Williams stroke that lofted Rip Sewell's "Eephus" pitch over Fenway Park's right-

Condition	Change in Distance	Home Run Probability
1000 feet of Altitude	+7 feet	+12%
10 degrees air pressure	+4 feet	+7%
10 degrees ball pressure	+4 feet	+7%
1-inch drop in barometer	+3 feet	+5%
1-mph following wind	+3 feet	+5%
Ball at 100 percent humidity	−30 feet	−50%
Pitch, + 5 mph	+3.5 feet	+6%
Hit along foul line	+11 feet	Not defined
Aluminum bat	+35 feet	+60%

TABLE 5.1: *Distance added to 400-foot fly ball hit to center field by changes in conditions and corresponding proportional changes in home run probabilities*

field fence in the 1946 All-Star Game probably generated little backspin, Williams was still able to hit the ball over 340 feet. The Eephus pitch—tossed about 25 feet in the air to drop across the plate at a considerable angle—is seen today mainly in softball. Aside from losing the 20-foot flight advantage of backspin, Williams had to provide almost all of the power—equal to an extra 50 feet or so of flight—to drive the 30-mph pitch over the fence. But only Babe Ruth could bat like Ted Williams, and no one else ever hit a home run off Sewell's blooper.

Batters sometimes work on their bats (illegally, but this is only a venial sin), grooving the surface to add backspin. However, it's unlikely that there is any benefit in this beyond occupational therapy. For all of the small-angle impacts that lead to fair balls, the normal ball-bat friction is sufficient to ensure maximum backspin. Even as good, well-grooved tires do not stop a car faster than smooth, bald tires as long as the car doesn't skid, fair balls hit off a smooth bat probably don't skid and then spin as much as those hit off a grooved bat. However, grooves are important on lofted golf clubs, like the 9-iron or pitching wedge, where hard steel, lubricated by the moisture from mashed grass, meets the smooth, enameled surface of the golf ball at a large angle. But they are not important for the less lofted clubs such as the long irons and woods, where the squarely hit ball has less inclination to skid.

The inclination of the arc of the swing does not strongly affect the velocity of the struck ball, but it does affect the mean angle at which the ball leaves the bat and, hence, the probability of hitting a very long ball and a home run. The great high-average line-drive hitters—such as Rod Carew, Wade Boggs, and Tony Gwynn—swung the bat such that the barrel crossed the hitting region just in front of home plate, traveling upward on the same line that the average pitch is moving down—that is, at an angle of about 8 to 10 degrees. (Players call this a level swing.) Then, if their bat position is correct but their timing is slightly off in their effort to hit the ball over second base, they still connect with the ball squarely; if they are early by 9 inches on the fastball they (all left-handed batters) will pull the ball in a line drive between first and second into right field; if they are equally late they will hit a line drive to the opposite (left) field. If their perfectly directed swing hits the ball absolutely squarely, the result will be a line drive a few feet over an infielder's reach. If they hit the ball half an inch high, they will hit a very hard ground ball that will often go through the infield; if they are half an inch low, they will hit a high line drive for a double if it goes in the gap between fielders—or down the line. But they will hit few home runs—and very few pop flies. They hit few home runs, not because they do not hit the ball hard, but because they do not hit the ball up. Boggs, in particular, hit the ball very hard and won home run contests against certified sluggers by swinging up a bit on the ball.

The home run hitters with relatively low batting averages but high slugging average uppercut the ball, typically swinging in such a way that the bat is moving upward at an angle of as much as 20 degrees as it meets the ball. Reggie Jackson, who hit a lot of home runs circa 1980, seemed sometimes to swing up at an even greater angle. Roughly speaking, every 10 degrees extra—i.e. beyond 10 degrees—of uppercut sends the squarely hit ball at an upwards added angle of about 16 degrees. Therefore Jackson's squarely hit ball from a 20-degree swing took off going upward at about 26 degrees and Carew's solid hit came off his 10-degree swing moving up at an angle of 10 degrees. But for equally hard hit balls, Carew ended up sliding into second with a drive that landed about 210 feet from home plate and skipped between the outfielders, while Jackson took a leisurely tour of the bases with the ball in the pocket of a fan who caught it 375 feet away in the right-field bleachers. Of course, if Jackson, the bigger man, actually hit the ball a little harder, it would go even farther.

But a mistimed swing would have had different consequences for these two kinds of batters, as suggested by the diagrams of Figure 5.11. If they missed their timing by swinging 9 inches too early on the fastball, Carew, batting left-handed, would still hit the ball squarely, but now on a line over the first baseman's head. However, Jackson, also batting left-handed, would top the ball and run out a weak grounder to the first baseman. If they swung

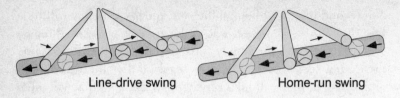

Line-drive swing   Home-run swing

FIGURE 5.11: *Ball and bat trajectories showing the consequences of a mistimed swing by a left-handed batter with a level swing like Rod Carew, who would hit the ball squarely despite an early or a late swing, and by a batter with an uppercut like Reggie Jackson, whose too-early swing wound up under the ball and his late swing over it.*

equally late, Carew's squarely hit ball would pass in a line over the infield between short and third while Jackson, hitting under the ball, would hit a high pop fly to be gathered in by the third baseman. Hence, Reggie got the home runs while Rod hit for the high average. And both are on Bill James's list in his *Historical Baseball Abstract* of the 100 greatest who ever played the game.

HOW FAR CAN A BALL BE HIT?

Stories of colossal clouts, of balls hit tremendous distances, constitute a part of the apocrypha of baseball. The stories of the longest blows (carries of from 550 to 600 feet are usually reported) always place the ball's landing in some poorly surveyed location—e.g., the street outside the stadium—so that "estimate" might best be translated as "uninformed wild guess."

Although the question as to how far a baseball can actually be hit is not easy to answer definitively, we can make some useful comments about the issue. It is surely not easy to hit a baseball 450 feet. But balls have been *caught* about that far from home plate in the deep center fields of the old Yankee Stadium and the New York Giants' Polo Grounds. At Yankee Stadium, Joe DiMaggio pulled down a Hank Greenberg drive in 1939 that traveled nearly 450 feet. And Willie Mays ran almost to the clubhouse stairs, about 475 feet from home plate in the old Polo Grounds, to catch a towering Vic

Wertz smash with two men on to break the hearts of the Indians in the 1954 World Series. Even with the hyperbole removed from a catch that didn't need it, the ball probably traveled nearly 450 feet.*

And there are home runs that we know have traveled nearly 500 feet, where the distance is well defined simply by the balls' direction and by the fact that they *were* home runs. Phillip Lowry, in his book on ballparks, *Green Cathedrals,*† notes that Joe Adcock, Lou Brock, Henry Aaron, and Luke Easter (in a Negro League game in 1948) all hit home runs over the nine-foot screen in front of the center-field bleachers 480 feet from home plate in the Polo Grounds. On the side of limits, some interesting, and well defined, negatives are known. For example, no one—not Ruth, not Gehrig, not DiMaggio, Mantle, or Maris— ever hit a ball out of Yankee Stadium.‡

We now have some interesting and accurate systematic information on the length of home runs from the IBM "Tale-of-the-Tape©" program, which was designed to determine how far home runs hit by the home team traveled (or would have traveled, if they had not landed in the stands). The scorer punched in the location where the ball landed (typically, the row and seat number or where it hit walls or screens) into an IBM personal computer loaded with coordinates from a survey of the park. The computer found the distance from home plate to where the ball landed and the height of the landing point above the playing field. Then it added a length derived from that height to come up with the distance the ball would have traveled on a level playing field. The algorithms used were quite good, and the final computed distance was probably accurate to better than 10 feet.§

---

*These balls would have stayed in the air about 6 seconds, and that is time enough for a Mantle or Mays to run 150 feet under the ball.

†Addison-Wesley, New York, 1992.

‡However, some say that Josh Gibson, the great catcher of the Negro Leagues' almost legendary Homestead Grays, did hit one out.

§IBM made a contribution to a charity designated by the team in the name of those who hit the longest home runs.

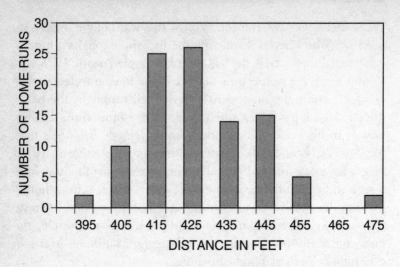

FIGURE 5.12: *The distribution of lengths for the three longest home runs hit at home by 15 major league teams in 1988 and by 19 in 1989.*

Figure 5.12 shows the frequency distribution of the length of the three longest home runs hit during 1988 and 1989 by home team players from the teams participating in the IBM program (15 clubs in 1988 and 19 in 1989, of the 26 major league teams). Of about 2,000 home team home runs, only two carried over 460 feet—a 478-foot blow by Dave Winfield and a 473-foot home run by Fred McGriff. Only eight, or about 1 in 250, traveled more than 450 feet.

So long home runs are surely rare, but if in two years, there are only two hits that traveled about 475 feet, what is the record? Dan Valenti, in his book about famous home runs, *Clout,** describes what might reasonably be considered the longest home run for which a reliable distance can be established. Mickey Mantle, batting right-handed, hit a Chuck Stobbs pitch out of Griffith Stadium in 1953 that a Yankee publicity man claimed traveled 565 feet. Mantle's blow surely didn't go that far in the air, but it did go a long way, and we can estimate how far rather accurately.

The ball was seen to glance off a beer sign in left-center field as it cleared the 55-foot left-field bleachers 460 feet from home

---

*Stephen Greene, New York, 1989.

plate; the spot at which it hit the beer sign was 60 feet above the playing field level. In general, balls go farthest when hit at a launching angle of about 35 degrees; with a strong following wind, the optimum angle is more like 40. degrees. And—as a consequence of air resistance—a ball comes down at an angle greater than that at which it takes off. If the ball came down at an angle of 50 degrees, it would have hit the ground 510 feet from home plate; a more precise calculation gives an answer of 506 feet, with an uncertainty I put at no more than 5 feet.

A probable trajectory of the ball Mantle hit is shown in Figure 5.13. It is obvious from the scale drawing that the claim of 565 feet is nonsense, but that an estimate of 506 feet is quite reasonable.

When Mantle hit Stobbs's pitch, there seems to have been a strong following wind and I'd expect that the longest home runs hit in outdoor parks are always wind-assisted. Valenti quotes Sam Diaz, a meteorologist who was working that day in the Washington Weather Bureau, to the effect that the wind during that period was 20 mph, with gusts up to 41 mph. If we take the wind as blowing out at 20 mph 60 feet above the playing field, but shielded by the stands below 60 feet, we can conclude that a ball hit that hard would carry about 430 feet on a windless day; and if the wind had been *against* Mantle, the ball would have been only a routine fly out.

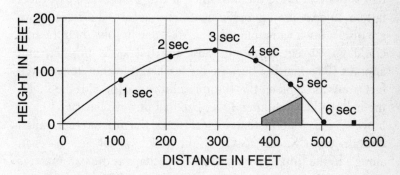

FIGURE 5.13: *A plausible trajectory of the ball hit by Mickey Mantle off Chuck Stobbs that went over the left-field bleachers at Griffith Stadium. The solid square shows the distance the ball was claimed to have traveled.*

But major league baseball had been played in Griffith Stadium for almost fifty years, and the wind must have been blowing out often, and therefore, since no one else ever hit a ball that left Griffith during a major league game,* Mantle's feat shouldn't be denigrated.

Although the question as to how far a baseball can actually be hit must eventually be answered through experiment, it is possible to make some useful comments on the issue. First, though many balls are hit 400 feet, such a blow constitutes quite a long home run. According to the graph in Figure 5.8, the bat speed required to hit an 85-mph fastball 400 feet to dead center field under standard conditions (at sea level, no wind, moderate temperature, and so forth) would be about 70 mph. To hit the ball 450 feet, the bat must be swung at a velocity of 80 mph, 14 percent faster with an energy 30 percent greater; and for the ball to go 500 feet, 65 percent more energy is needed. As this is a very large difference, we can estimate that 450 feet is about the maximum that ballplayers can hit the ball *under standard conditions.*

Although 450 feet may be the maximum distance a ball can be hit under normal conditions in Old New York, baseball is played under other conditions and other places. Let us say that a pitcher is very fast and that his ball velocity is 95 mph. This increases the distance the ball would be hit from 450 to about 455 feet. If the ball is pulled toward the foul lines, it will go as much as 15 feet farther, and we are up to 470 feet. Let us assume further that we are discussing a game played on a 100° day in July; then the ball could go 10 feet farther because the hot air is thinner, and another 10 feet farther because the hot ball is livelier, to land 490 feet from home plate. If a 10-mph following wind aids the flight, the ball could go another 40 feet, so that now we have a 530-foot blow. (We exclude meteorological freaks; the tornado that wafted Dorothy from Kansas to Oz could well have taken a baseball along.) In the thinner air of Atlanta—perhaps as the barometer is

---

*Josh Gibson is said to have cleared that wall *twice* when the Homestead Grays played home games at Griffith Stadium in the 1940s.

falling, with lightning on the horizon—the ball might go another 10 feet, to give us 540 feet. And in Denver, this distance could reach 570 feet!

In summary, if someone tells me that a ball was hit 550 feet anywhere in the majors but Denver, I won't believe it—but I won't bet the farm against it. And if the ball is hit in Denver, all bets are off—it might nearly reach Kansas, or even Oz.

In my estimates of maximum-length home runs, I have limited myself to home runs *by major league baseball players in games.* If a sufficiently tempting prize were given for the ball hit farthest under baseball rules, one might recruit an Olympic-level weight man, quick as a cat and carrying near 300 pounds on a near 7-foot frame, to do the job. Swinging a 42-inch-long, 4-pound bat after taking two steps—like the young Babe Ruth—into fastballs thrown over the center of the plate, I estimate, he could hit the ball well over 500 feet under standard conditions at sea level.

If a player can hit a ball 550 feet—maybe—how far can one throw a ball? In Omaha in 1957, Glen Gorbous threw a ball 446 feet. Assuming he threw on a hot day with a 5-mph following breeze at the Omaha altitude of 1040 feet, I estimate that Gorbous must have thrown the ball with an initial velocity of about 115 mph. In Chapter 4, I noted that the very fastest pitchers throw with a muzzle velocity of about 107 mph. But pitchers throw from a stationary position, while Gorbous threw from a run, which is sufficient to account for the extra velocity.

THE NUMBER OF HOME RUNS

Just as players and fans are especially interested in how far home runs can be hit, they are interested in how many home runs are hit.

Baseball was originally a meadows game played without fences. When it was discovered that people would pay admission to watch, it became necessary to fence in the game, but to preserve its character the fences were usually placed far away to keep them out of play. As in the meadows game, major league home runs in the early days were largely in-the-park home runs—just stretched triples. And before WWI, there were far more triples hit in the major leagues than home runs. Players did not try to hit the ball over the fielders' heads—for most players that was only a recipe for a loud out. The line drive was king, and everyone tried to hit the ball squarely and avoid high fly balls.

But then, as now, home runs were valuable. Then, as now, a home run was worth about 1.4 runs on the average (when men on base are considered), and then, as now, teams scored between 4 and 5 runs per game on the average.

However, over the period from 1910 to 1920, the major league teams all built urban ballparks, where real estate was expensive, and the fences were closer and thus came into play. Players, coaches, managers, and owners still knew the game as a meadows game, and the players—with support from coaches, managers, and owners—continued to aim to hit line drives. Then along came a big left-handed pitcher, batting ninth in the lineup and, as a pitcher—and a great one—under no pressure to hit well. The kid loved to swing the bat, he swung from his heels, and with reactions and coordination that a tiger would have traded a hind leg for, he regularly hit the ball over the wall. Recognizing the value of his long home runs, the Boston Red Sox converted Babe Ruth from a one-in-four-day pitcher to an everyday outfielder, and his home runs changed baseball forever.

In 1919, his first full year as an outfielder and batter, Ruth hit more home runs (29) than all but five teams in the two leagues. The next year only one team (Philadelphia, in the National League, hit 60) hit more home runs than Ruth's 54. The other players were still swinging to hit line drives, but soon enough some of them began swinging for the fences. Rogers Hornsby, who hit 9 home runs when he won the National League batting title in 1919 by hitting .370, decided that he, too, could hit home runs, changed his batting style, and hit 42 home runs in 1922 (while compiling a. 401 batting average), giving him even more than Ruth, who hit 35 that year. Though few of the older players could adjust, younger men began to swing as much like Ruth as they could. Hence, as shown in Figure 5.14, it took only a few years to take the average number of home runs^d per game (by both teams) from about 0.3 in the second decade of the century to a level of about 0.75.

As the tactical value of the home run became more obvious, the team front offices, their managers, and their players began to direct their play more toward the home run and the number of home runs per game (by both teams) increased steadily from about 0.75 in the 1920s to nearly 2.0 in the late 1950s. However, since most players who swing for the fences miss *everything* a lot, strikeouts per game (shown in Figure 5.15) increased correspondingly, from about 6 in the 1920s to nearly 12 in 1965. Both home run and strikeout frequencies dropped off during the following years—for what reason I know not. But both increased in the 1990s, so that at the beginning of the third millennium players hit more home runs—and strike out more—than ever before.

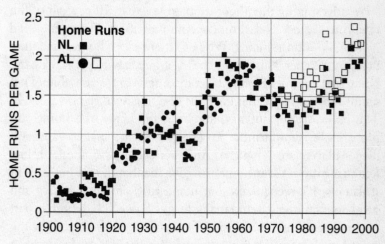

FIGURE 5.14: *The average number of home runs hit per game in the National and American Leagues from 1902 to 1998. The open squares show American League totals after 1973, when the designated hitter was introduced, increasing the number of home runs per game by more than 10 percent.*

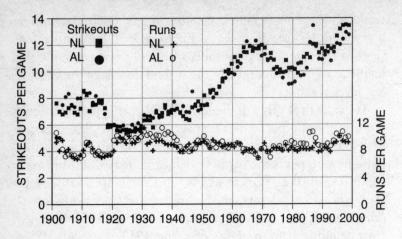

FIGURE 5.15: *The average number of strikeouts per game in the National and American Leagues from 1902 to 1998, plus the average number of runs scored per game.*

Of course, the radical increase in home runs has changed the game. Before WWI, home runs (including the men on base who were driven in by them) contributed less than 10 percent of the total runs scored. Today, they account for about 35 percent.

The question remains, Why are there more home runs than there used to be? This is more fun to argue than to pontificate on, but some comments might be in order—and excusable. 1) It seems most unlikely that the balls have changed significantly. 2) The ballparks are not so different as to change the home run probabilities significantly. 3) The players are bigger—and still bigger players are sought. 4) And, perhaps most important, batters swing more to hit home runs; we see that in the increases in strikeouts. It's recognized that home runs are good tactics and good entertainment—and major league baseball is definitely part of the entertainment industry. Thus, home run hitters draw the great big paychecks while singles hitters have to make do on a paltry million dollars a year or so.

TECHNICAL NOTES

a) For a given velocity of impact, the area under the lower curve in Figure 5.4 is proportional to the change in velocity of the ball.

b) I use the following formulae to calculate the trajectories of batted balls:

$$\nu'_n = C\nu_n \quad \nu'_t = \sqrt{5/7} \cdot \sqrt{0.5C + 0.5} \; and \; \omega = \nu'_t/r$$

where C is the coefficient of restitution, ν_n and ν'_n are initial and rebound normal velocities in the ball-bat center-of-mass system (which is equivalent to the ball-wall system of Figure 5.2), ν_t and ν'_t are the associated tangential velocities, ω is 2π times the rate of spin of the ball (in revolutions per second) and r is the ball radius. The factor $\sqrt{5/7}$ accounts for the transfer of translational kinetic energy to rotational energy, and the factor $\sqrt{0.5C + 0.5}$ is introduced as an estimate of the reduction of the transverse rebound velocity that might follow from frictional effects. These recipes are designed to represent collisions at an angle less than 30° from the normal (which includes all hits). In the case of more nearly glancing collisions, the ball may skid off the bat, losing little transverse velocity and picking up little spin.

c) My colleague R. C. Larsen put several balls in a deep freeze at about −10° F and several in an oven at about 175° F for roughly a day. He then let the balls sit at room temperature for an hour (so the covers seemed normal) and measured their coefficient of restitution (COR) at a velocity of 25 mph. As one might guess, the cold balls were dead—their COR was reduced about 10 percent from normal—and the heated balls were quite lively, their COR increasing about 13 percent. This translates into a change in the distance of a fly hit 375 feet with a normal ball to about 350 feet for cold balls and over 400 feet for hot ones.

d) I find the year-to-year difference in the number of home

runs puzzling. It is useful to discuss the variations in terms of pure chance—as if the probability of hitting a home run in a given at-bat is the same for every player and about equal to the probability of a Monte Carlo roulette wheel coming up 17 (or any other number). Based on this assumption, the distribution of frequencies of home runs would follow a *normal distribution* curve, or "bell curve," with the center the "true" average and with a width about equal to the square root of the average number of home runs—if the "true" average were 1,600 home runs, we would expect that about two-thirds of the time the number would be between 1600–40 and 1600+40.

For real baseball seasons, I would expect a somewhat larger spread owing to specific chance factors—such as the year-by-year variation in the number of good home run hitters in each league—but the spread or width from the simple model is a useful base.

Using this logic, I find the difference between the number of home runs hit in the two leagues each year to be about 1.5 times that which was expected from the simple roulette-wheel model. This seems quite reasonable. But the year-to-year variations are much larger. The differences between the number of home runs hit in a given year and the average of the years before and after are about 2.5 times chance. This is significant.

In short, the number of home runs hit in a year for each league fluctuates greatly—but the numbers hit in the two leagues are correlated. When the number of home runs hit in the National League goes up, the number hit in the American League tends to go up too. Why the correlation? I don't know. The only obvious suspects are the weather and the elasticity of the balls. A hot year would increase the number of home runs for each league. But year-to-year temperature differences, averaged over all ballparks, seem far too small to account for the correlations. Also, the two leagues use the same balls. If the elasticity of the balls changed year to year, the number of home runs hit would tend to vary, and vary the same way in both leagues. But, again, the variations seem much too large to attribute to that cause.

In summary, I don't understand the statistics of home runs, and thus I tend to respond to all highly technical statistical analyses of baseball with Mark Twain's comment, "There are lies, damn lies, and statistics."

6

PROPERTIES OF BATS

THE OPTIMUM BAT WEIGHT

From the *Official Baseball Rules*:

1.10 (a) The bat shall be a smooth, rounded stick not more than 2¾ inches in diameter at the thickest part and not more than 42 inches in length. The bat shall be one piece of solid wood.

NOTE: No laminated or experimental bats shall be used in a professional game (either championship season or exhibition game) until the manufacturer has secured from the Rules Committee approval of his design and methods of manufacture.

(b) Cupped Bats. An indentation in the end of the bat up to one inch in depth is permitted and may be no wider than two inches and no less than one inch in diameter. The indentation must be curved with no foreign substance added.

(c) The bat handle, for not more than 18 inches from the end, may be covered or treated with any material to improve the grip. Any such material or substance, which extends beyond the 18 inch limitation, shall cause the bat to be removed from the game.

No colored bat may be used in a professional game unless approved by the Rules Committee.

In fact, the bats used by players are almost never longer than 36 inches. Once they were made of hickory and weighed from 45 ounces upward to the 56 ounces of lumber supposedly wielded by the young Ruth,* but today's bats are made of ash and range from 31 to 35 ounces for the most part. The typical dugout bat rack in 1920 probably held no bat lighter than 36 ounces; the 2001 bat rack probably holds no bat heavier than 36 ounces. But the choice of weights and shapes has always varied considerably from player to player.

If we understand the mechanics of batting, it might seem, we can establish the optimum size and weight of the bat based on simple physical principles. However, a number of factors suggest that we cannot expect to reach such definite conclusions. Most important, the broad choice of bats used successfully by baseball players tells us that success in batting cannot depend sensitively on the character of the bat. The sensitivity of the game to small differences in batting means that precise analyses may be required if reliable conclusions are sought concerning those small differences (e.g., there is only about a 5 percent difference in the length of flight of the average home run and a long fly ball caught on the warning track), and the act of batting is, in itself, complex and not precisely understood.

At this time I will describe my analyses solely in terms of the weight of the bat, since this—along with its length—is a commonplace parameter. I'll then neglect differences in the distribution of the weight along the bat (i.e., the shape of the bat), though that can be important; obviously, the addition of an ounce in the

---

*Hack Miller, Cub outfielder of the early '20s, was said to have swung a 69-ounce bat!

handle of the bat has much less effect on the swing than the same ounce added to the barrel. Also, I am neglecting tactical differences; we might reasonably expect that line-drive hitters like Tony Gwynn might want bats of a different character than would home run specialists like Mark McGwire. However, in our analyses, I'll emphasize home runs, as home runs are important and less dependent upon tactical decisions than lesser blows.

I will make the simplifying assumption that the weight and weight distribution of a bat can be changed without modifying the shape of the bat. This can only be entirely valid if a complete range of wood densities is available, and that is not the case. Baseball bats seem to have been made only of hickory or ash; the heavy bats used largely in the first decade of the last century were made of hickory with a density of about 0.82 times that of water; the light bats used today are turned from ash with a density of about 0.638. Ash is special; the American ash from which bats are made has an unusually high strength-to-weight ratio. A hickory bat with the same dimensions as the 33-ounce ash bat used by Roger Maris to hit 61 home runs would weigh about 42 ounces. The 48-ounce hickory bat swung (choked up) by Edd Roush to give him a lifetime average of .323 between 1913 and 1931 and the 47-ounce hickory bat swung (held at the end) by Babe Ruth to hit 60 home runs in 1927 had larger barrels and thicker handles than Maris's bat and were probably an inch or so longer.

The rules of baseball allow a barrel diameter of 2.75 inches. In order to keep the bat weight down, the light bats used today generally have smaller barrels—typically 2.50 inches in diameter. This reduces the bat-ball hitting area (which is proportional to the sum of the diameters of ball and bat) by about 4 percent and the weight by about 4 ounces for an ash bat. The heavy bats, usually 38 ounces, used by Mickey Mantle and the 40-ounce bat swung by Dick Allen had full 2.75-inch barrels. Of course, longer bats weigh more; an inch of extra length in the barrel adds nearly 3 ounces to the weight of a hickory bat and a little more than 2 ounces to an ash bat, though an inch added to the handle adds only about a third of an ounce.

The bats of the great home run hitters varied in weight over a broad spectrum. In 1918, early in his career—when he led the American League with 11 home runs, though he was used mainly as a pitcher and spent only about a third of the season in the outfield—Babe Ruth was reputed to have sometimes used a 56-ounce hickory bat. In 1927, when he hit the record of 60 home runs in 154 games, he seemed to have favored a 47-ounce Hillerich & Bradsby slugger. Later he dropped down to a 44-ounce bat, and in his last year, with the Boston Braves—when he quit early in the season with only 13 hits in 72 at-bats, though 6 of those hits were home runs—he was said to be experimenting with bats as light as 36 ounces. And Ruth's last home run, the last he ever hit, was the first ball ever hit to right out of Forbes Field in Pittsburgh! But most players used much lighter bats then, as they do now. Roger Maris hit 61 home runs with a 33-ounce bat and Hank Aaron used 31- and 32-ounce bats to hit more major league home runs than anyone else. Though Ernie Lombardi used a 42-ounce bat in the 1940s, no other major batter since Ruth seems to have hefted more than 40 ounces.

If one player can hit a significant number of home runs with a 56-ounce bat and another can be equally successful with a 32-ounce bat, one is led, inevitably, to the conclusion that the distance a well-hit ball travels cannot depend dramatically on the weight of the bat. Furthermore, since there are obviously advantages to a light bat since it must be easier to time a pitch with the quick swing possible from a light bat, there must be compensating advantages to a heavy bat, or Ruth and Lombardi would not have used heavy bats.

The heavier and larger bat, with its larger and longer barrel and longer good hitting length, provides more good hitting area, but such a bat probably also transfers a higher velocity to the well-struck ball; it seems likely that a player can actually hit the ball a little farther with a heavy bat.

But could most players choose badly? Could an analysis of the physical principles of batting lead us in a different direction? Because of the difficulty of obtaining a deep understanding of so complex an activity as batting, we must begin simply. From elementary principles of mechanics, we can say with complete reliability that for a given bat speed, a heavy bat will drive a ball farther than a light bat. Conversely, for a given kinetic energy of the bat, a light bat will drive a ball farther than a heavy bat (for bat weights greater than 20 ounces). To this we add the (very plausible) condition that no player can swing a heavy bat faster than he could a light bat. And we hold that no player can put more energy into a light bat than into a heavy bat. The energy transmitted to the bat is simply the product (better, *integral*) of the force to the bat that the hands apply along the direction of motion times the distance through which the hands move. If we assume that the arc of the hands—and bat—is the same for a light or heavy bat, the larger force that it is possible to apply to the more slowly moving heavy bat will result in a larger energy transfer to that bat.

Figure 6.1 shows the variation of maximum distances with bat weights that a 90-mph fastball can be hit under these two conditions; the dotted line shows the variation if bats of different weight are swung with the same kinetic energy that drives a ball 381 feet off a 32-ounce bat; the broken line shows the variation if the bats are swung with the same velocity as that of the 32-ounce bat. Noting that these curves establish *limits* on the variation of distance with bat weight, we can be reasonably confident that if a man who hits a ball 381 feet with a 32-ounce bat changes bat weight, and swings just as hard, the distance the ball will go will fall between these two extreme lines. Moreover, it is clear that the man who hits 381-foot home runs with a 32-ounce bat will not likely gain or lose very much distance by changing the weight of his bat by a few ounces.

Although this is as far as physics can carry us reliably, it is interesting to investigate more detailed results from a simple but plausible model of bat and batter. In this model, we assume that the

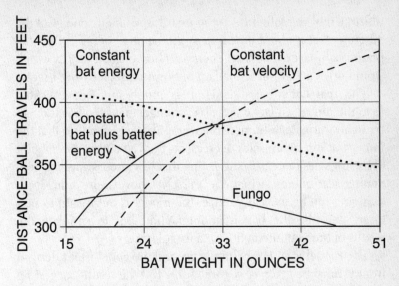

FIGURE 6.1: *The curves show the distances a 90-mph fastball can be hit by bats of different weights swung at a definite velocity, swung with a definite energy, and swung such that the total bat-batter energy is held constant. The distance a stationary ball is hit (a fungo) by a similar bat swung with a definite bat-batter energy is also shown.*

bat and the body of the player move in the same general paths, driven by the same muscular forces, though different weight bats are used.[a] The forces generated by the muscles move the 180-pound batter just as they move the 2-pound bat. Since the same forces are applied over the same distances, the total energy generated to move bat and body during the swing is the same for different bat weights. If the bat is light, man and bat will move a little faster; if the bat is heavy, the swing will be a little slower. Fast or slow, light bat or heavy, the total kinetic energy of man and bat will be held constant if we assume a constant muscular force and a constant path of motion through the swing.

Though the energy is held constant for different bat weights, the energy is divided differently between bat and body. More of the energy goes into a heavy bat, though the bat—and player—velocities are smaller; less energy goes into a light bat, though the velocities of bat and player are somewhat greater. In the case illustrated by the solid line in Figure 6.1, the kinetic energy of the 32-ounce bat is about the same as the kinetic energy of the 180-pound ballplayer. The total energy is chosen again as that which would drive a 90-mph fastball an optimum distance of 381 feet.

The real batter-bat interaction is much more complicated than the model. In fact energy is exchanged between bat and batter in an intricate fashion. Moreover, physiologists tell us that the forces put out by muscles decrease as the velocity of contraction of the muscles increases, suggesting that less force is applied, generating less energy, when the muscles move more quickly in swinging the lighter bat. Hence, the model is not meant to precisely describe the batter-bat interaction but to simulate the results of the real interaction in a useful way.

The model is certainly simplistic, and the conclusions derived from it must be taken *cum grano salis*. But the results suggesting that the maximum distance of 389 feet would be achieved if the batter used a 46-ounce bat are interesting and not implausible. The effects are not large: The 46-ounce bat propels the ball only 8 feet farther than the 32-ounce bat; if the batter drills a hole in his 32-ounce bat (perhaps filling it with cork) so its weight is reduced to 28 ounces, the ball will go only 373 feet.

Armed with the model (albeit simplistic), one can look at other factors. How does the optimum bat weight vary with the size of the player? Babe Ruth was reported to weigh 251 pounds in 1927 when he hit 60 home runs. According to our model, the Babe would have been best suited to hit 90-mph fastballs for distance with a 52-ounce bat. Conversely, Lloyd Waner probably weighed little more than 140 pounds at times. Our model tell us that "Little Poison" would hit fastballs hardest with a 42-ounce bat. Using the approximation that the total energy output of a ballplayer is proportional to his weight, we can estimate the extent to which the length a player can hit a ball depends on his size. Hence we find that the same proportionate effort that allowed 140-pound Lloyd Waner to drive a 90-mph fastball 338 feet with a 32-ounce bat would allow 180-pound Ducky Medwick to hit the ball 362 feet using a bat of the same weight. And if

225-pound Mark McGwire swung equally hard, his shot could land in the bleachers 382 feet from home plate.

If the ball is moving more slowly, the optimum bat weight declines. The 180-pound model player would best use a 48-ounce bat to reverse the velocity of a 95-mph Roger Clemens fastball and send it toward the center-field fence. But a 42.5-ounce bat would better drive the 80-mph slider, and the 65-mph breaking ball would be hit farthest by a 37-ounce bat. And if that 180-pounder wanted to hit fungoes (or bat a ball off a tee), he would find that a 24-ounce bat would do best—if he could find one. Indeed, special fungo bats are made to be quite light by reducing the diameter of the barrel.

By and large, these results are in accord with experience and common sense. However, except for the fungo bat, all of the optimum bat weights seem rather high, even if they are optimized for slower velocities than the major league fastball. But to survive in professional baseball, one must hit the ball often as well as hard. The model player would lose only a little distance by going to a 32-ounce bat, and with the extra quickness afforded by the lighter bat, he could expect to make good contact more often.

Have any measurements been made that might substantiate the claims of the simple model that heavy bats drive the ball farther? Yes! Many years ago, on May 20, 1962, *This Week* magazine reported the results of an experiment in which Roger Maris batted for distance with five different bats that varied in weight from 33 to 47 ounces. The bats were copies from Hillerich & Bradsby records of the bats of previous great home run hitters (Babe Ruth used a 47-ounce bat to hit his 60 home runs). The pitcher was a veteran batting practice pitcher from the Yankees who expertly served up hittable pitches. Though there were fluctuations in the length of the drives, of course, a statistical analysis confirmed the trend seen casually: The heavier the bat, the farther Maris hit the ball. Indeed, the best fit to the data showed an increase of range with bat weight about the same as that shown by the dotted constant-velocity line of Figure 6.1, though that limit was within the uncertainties of the data. Maris probably adjusted his swing so that he swung the heavier bats almost as fast as he did the lighter bats; perhaps he lengthened his swing when he used the heavier bats to hit the batting practice pitching. However, although the average length of balls he hit with his own 33-ounce bat was the shortest of all, Maris returned to that bat to face hostile pitching.

If the loss in distance that follows from a choice of a lighter bat is not great, is the gain in time in swinging the lighter bat substantial? It is easy to set limits on the difference in timing using the extreme model of the constant-energy bat. If we assume that the batter's "time of decision" is typically 150 milliseconds before the ball crosses the plate, we can conclude with some assurance that the gain in the time required to swing the bat that follows from the change from a very heavy 38-ounce bat to a moderately light 32-ounce can be no greater than 13 milliseconds, or about 1'9", in the flight of the 90-mph fastball. Our more realistic model cuts that down to about 10 milliseconds and 1'3". Hence, the batter would gain a little over a foot on the fastball by shifting from a 38-ounce bludgeon to his favorite 32-ounce bat. If he drills a hole in that 32-ounce bat, reducing the weight at the barrel by about an ounce and a half, he would gain about 5 milliseconds, or about 6 inches on the fastball. While this gain in time is small compared with the 425-millisecond pitcher-to-batter time for a good fastball, from the discussion in Chapter 3 we see that the time should better be compared with the 125 milliseconds that are left for gathering and processing information after subtracting the 300 milliseconds required to process the initial visual information and swing the bat. Hence, for some players, the addition of 5 or 10 milliseconds to that 125 milliseconds could well be more important than the small loss in the speed—and flight distance—of the ball as it comes off the slightly lighter bat.

Such a simplistic, essentially kinematic argument ignores the athletic complexity of the swing of the bat. The fine batter's "beautiful" swing that supplies maximum energy to the bat as it strikes the ball follows from precise timing of the various actions that contribute to the swing. It seems possible that for a given batter, a slightly lighter—or slightly heavier—bat may lead to a better (-timed) swing and a larger energy transfer to the bat. And this could well override any pedantic kinematic derivation of an ideal bat weight.

Though these discussions define the considerations applicable to choosing an optimum weight of bat—and support the commonsense view that players now choose bat weights sensibly—is it not possible that a better weight distribution would serve to drive a ball faster and farther? Could the standard bat be badly designed?

The kinetic energy supplied by the player to the bat can be divided usefully into two parts: an energy associated with the linear motion toward the ball by the mass of the bat, and an energy associated with the rotation of the bat about the point of impact with the ball. This rotational energy, which is typically about 5 percent of the total energy, does not contribute to the driving of the ball. If a bat could be designed such that almost all of the mass were concentrated very near the point of impact—a bat with a short, heavy barrel and a handle like the wooden shafts of Bobby Jones's golf clubs of the 1920s would do the trick—the rotational portion of the bat energy would be much reduced. Such a bat, swung with the same total energy, would hit the ball somewhat farther; the ball hit 400 feet with the standard bat would go almost 410 feet off the golf-club bat. But the small increment of extra distance would be achieved at the cost of a considerable reduction of the region of good hitting and a considerable increase in the cost of broken bats. A ball hit a few inches from the optimum point along the axis of the bat would probably not leave the batting box much faster than the broken-off head of the bat.

HITTING THE BALL IMPERFECTLY

This discussion of batting has so far considered only almost perfectly hit balls. Since it is a pitcher's goal to see that balls are not well hit—and pitchers are sufficiently successful to hold mean batting averages well below 30 percent—mishit balls are the rule rather than the exception. The bat must be swung with great accuracy with respect to the trajectory of the pitched ball if the ball is to be well hit. It has been noted that the swing must be timed correctly, with an error no larger than 1/100 of a second, if the ball is to be hit fairly somewhere between first and third base.

Rather precise vertical orientation is also required. Figure 6.2 shows the trajectories of balls hit by a bat with a 2.5-inch-diameter barrel with different bat-ball vertical displacements. I assume that the swing is otherwise well directed and that the bat is swung with a speed of 65 mph upward at the Rod Carew line-drive angle of 10 degrees at an 85-mph overhand fastball, that crosses the plate dropping at an angle near 10 degrees. The ball that is hit squarely is driven 200 feet as a line drive—rising to a maximum height of about 13 feet. If the bat is swung under the ball by 2 inches, such that the line of motion of the center of the bat passes under the center of the ball by 2 inches, the ball will go onto the roof of the stands behind the catcher. The ball hit 1 inch below center will be a routine fly ball out unless it is hit down the foul line, where it might be a home run. The ball will go far-

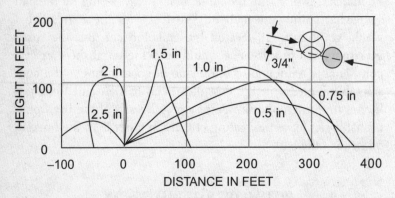

FIGURE 6.2: *The trajectories of balls struck by a bat swing under the ball by the given amounts. The bat speed is 70 mph directed upward by 10°; the ball speed is 85 mph directed downward by 10°.*

thest—about 350 feet—if hit about three-quarters of an inch below center. (The large backspin generated by the off-center bat-ball collision affects the pop-up trajectories noticeably.)

The paths of ground balls are more difficult to show, but excepting Baltimore chops, where the ball rebounds from home plate so high that the batter is on first almost before it comes down, the ground ball that goes for a hit must not be batted too sharply into the ground. By and large, for a ball to be hit such that it evades the fielders, the bat should strike the ball no lower than three-quarters of an inch below and no higher than three-eighths of an inch above center.

The batter is permitted a larger error along the axis of the bat. Though the nominal latitude of error of the position of the ball-bat impact point along the axis of the bat is greater than the allowable up-down error, the batter can adjust to changes in the vertical position of the ball more easily than he can to changes in the horizontal position. He does not need to change his stride or body orientation sharply to deal with the allowed vertical range of strike pitches, but he must change the position of his body to cover the 20-inch width of the strike zone. (The legal strike zone is a baseball-diameter wider than the plate, since the umpire is supposed to call a pitch a strike if any part of the ball passes through the strike zone.)

If the bat strikes the ball at some distance from the optimum point, the velocity of the ball off the bat is reduced. This loss follows from two different effects that we can treat somewhat independently. The velocity variation with position along the bat axis is determined by simple kinematic criteria defining the loss of energy from collision-induced rotation of the bat, and by less simple estimates of the loss of energy to vibrations of the bat along its length. I'll first discuss the purely kinematic factors by considering the ball-bat kinematics for a completely rigid bat. For expositional simplicity, it is useful to describe the reduction in energy transfer to the ball in terms of a loss of distance of a long drive to the outfield.

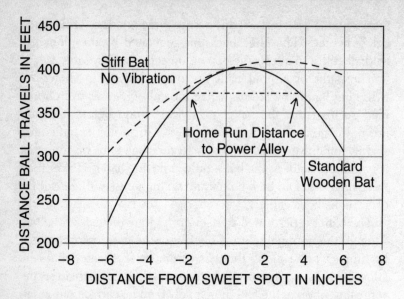

FIGURE 6.3: *The distance a ball can be hit by a 70-mph swing at an 85-mph fastball as the position of the bat-ball impact varies along the bat is shown for both a perfectly rigid bat and a typical wooden bat. The broken line marking typical distances to the fence at the power alleys suggests the typical home run distance.*

The broken curve in Figure 6.3 shows how the distance a ball travels changes as the point of impact is varied along a bat for a bat that is (unrealistically) so stiff that oscillations could be neglected. More realistically, I also neglect the clamping of the end of the bat by the hands of the batter.[b] For the particular bat under consideration, which is 34 inches long, the zero point is chosen to be the sweet spot, or the node of the fundamental vibration, about 27 inches from the handle and 7 inches from the end of the bat.

The force exerted on the hands by a ball mishit with a stiff bat is not too great. The ball's impact—which would severely sting the bare hands if caught—is greatly reduced by transmission through the bat. The maximum energy transfer from the stiff bat to the ball occurs when the ball is struck at a point about 5 inches from the end of the bat. Energy is lost in the impact of balls hit at the end of the bat, or near the handle, by the increased rotational energy the impact transfers to the bat. Hence, the variation in distance, reflecting a variation in velocity of the driven ball, follows simply from kinematic criteria.

But bats do vibrate when the ball is hit. The solid line in Figure 6.3 shows the distance the ball travels after being struck by a real wooden bat at varying distances along the bat barrel when the vibrations are properly considered. Hits near the optimal point shown in the figure cause minimum vibration. But hits far from that point generate strong vibrations in the bat, absorbing energy from the collision. Less energy to the ball means a weakly hit ball. And that vibration does sting the hands.

Under the large forces applied to the bat by the impact of the ball, the bat vibrates with different frequencies, each associated with a particular wave form. For each such wave form, there are points—nodes—that do not move. The center diagram of Figure 6.4 suggests the character of the lowest-frequency, fundamental vibration mode^c of a bat, and the upper diagram shows the nodes of the lowest three vibrational modes, which for a typical bat vibrate at frequencies of about 170 Hz, 570 Hz, and 1180 Hz. The energy given up by the bat-ball collision in bat vibration is largely stored in the vibrational energies of these three modes. Since the impact time of about 1 millisecond is larger than the vibrational period (the time for one full cycle) of the higher-frequency modes, those vibrations are not strongly excited and take up relatively little energy.

After the ball is struck, the motion of the bat is made up of the sum of the displacements of the different vibrational modes excited by the collision. The lowest diagram of Figure 6.4 shows the position of the bat 1 millisecond after collision with the ball, where the ball hits the bat about 2 inches from the fundamental vibrational node—170 Hz—in the direction of the handle. (The distortions are exaggerated in the diagrams; the maximum excursions are less than half an inch.) By this time the ball has left the bat. Note that the handle has not moved—the disturbance has not yet reached it.

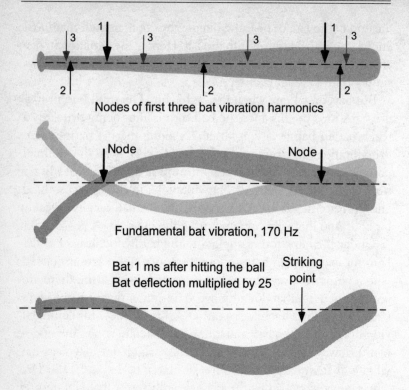

Nodes of first three bat vibration harmonics

Fundamental bat vibration, 170 Hz

Bat 1 ms after hitting the ball
Bat deflection multiplied by 25

Striking point

FIGURE 6.4: *At the top, the positions of the nodes of the three bat vibrations. At the center, the amplitude of the fundamental vibration. The node nearer the end of the bat is the baseball player's sweet spot. Shown below is the bat configuration 1 millisecond after hitting a ball 2 inches from the sweet spot, toward the handle. The amplitudes are exaggerated by a factor of about 25.*

When the ball is struck at the fundamental vibrational node, there is no stinging of the hands.[d] Ballplayers call this point the "sweet spot." Since the sweet spot is very near the striking point where the ball goes farthest (about 1 inch less measured from the bat handle), it is identified in the batter's mind with well-hit balls.

The magnitude of the induced vibration is proportional to the natural amplitude of vibration at the point of impact; hence, almost no vibrations are set up when the ball strikes the bat at a node, and maximal vibrations are set up if the point of impact is near an antinode. The energy of vibration is proportional to the square of the amplitude of vibration—twice the amplitude, four times the energy.

Since bats are not regular geometric objects, it is difficult to calculate their vibrational properties. However, L. L. Van Zandt has determined the vibrational modes of a standard wooden bat[e] and Allan Nathan has correctly calculated the effects of those vibrations on the transfer of velocity to the ball.[*] While the details of the calculations are not easily accessible, intelligent estimates of the amplitude of the vibrational effects can be made using *dimensional analysis*. The amplitude of the vibration induced by the collision of the bat and ball must be proportional to the impulse (the force applied to the bat multiplied by the time over which the force is applied), inversely proportional to the weight of the bat, and inversely proportional to the frequency of vibration. Knowing the impulse to the bat, and the weight of the bat, and having measured the frequency of vibration of the bat, I can calculate a typical amplitude of vibration induced in the bat by its striking a ball near an antinode to be a little less than half an inch; the bat will be distorted nearly as much as the ball. Using the same kind of estimate, I deduce that a significant fraction of the bat-ball collision energy can be transferred to the vibrational energy of the bat.

For the typical bat considered here, the frequency of the fundamental oscillation was measured to be about 170 cycles per second. Since the corresponding characteristic half-cycle time of about 0.002 seconds is appreciably longer than the natural bat-ball interaction time, which is about 0.0005 seconds (see Figure 5.4), the bat does not return the energy of distortion to the ball but retains it in that vibration familiar to any baseball player. That energy is, of course, lost to the ball, hence, the ball will rebound with decreased velocity.

The solid curve of Figure 6.3 represents an estimate I made,

*An account of Van Zandt's work was published in the *American Journal of Physics* in 1992. Nathan's work was published in the *American Journal of Physics* in 2000.

using Nathan's calculations, of the distance a batted ball travels as a function of the distance of the impact from the optimum point (which is near a node of the vibration) when the energy lost to vibration is taken into account.*

Although the half-period of vibration is sufficiently long compared with the time the ball is in contact with the bat that the energy is not returned elastically to the ball, the act of deforming the bat does lengthen the time of bat-ball contact (as the bat gives way at the ball's impact). The dominant sound produced in the course of a ball-bat contact is roughly half the inverse of the impact time (so that a collision that lasts less than $1/1000$ of a second generates sound frequencies greater than 500 Hz, about an octave above middle C). One hears these characteristic high frequencies in the "crack" of the bat. When the ball is hit off-center, the collision time is a little longer, generating somewhat lower maximum frequencies. With a lower maximum frequency and the addition of a strong component of lower-frequency sound from the natural bat oscillation, the "crack" becomes more of a "thunk." This sound is part of the pattern to which outfielders respond. A onetime center fielder advised me that "when the ball is hit straight at you (as discussed in Chapter 7, the distance such balls will travel is difficult to judge) if you hear the bat "crack," run back (toward the fence); if the sound is a "thunk," run in (toward second).

The curves of Figure 6.3 refer implicitly to a bat that retains its physical integrity through the collision. But we know that bats break, especially when the batter hits inside pitches near the handle. The phrase "sawed off the batter" is a metaphor with a bite. And when the batter is "sawed off" and delivers a broken-bat hit, the ball doesn't go very far—certainly not as far as implied by Figure 6.3.

We can understand this reduction in the energy returned to

---

*Trey Crisco and Michael Greenwald, working at the University of Rhode Island, made field measurements on batters hitting pitched balls that fit Nathan's calculations very well.

the ball in the course of the collision by the breaking of the bat, by considering the forces exerted on the bat by the ball's striking it near its handle. That collision bends the bat, stretching the wood fibers away from the ball and compressing the fibers on the side of the bat near the ball. The elastic response of those fibers results in an elastic resistance of the bat to the bending forces. The bending bat exerts a reaction force on the ball that increases as the bat bends; indeed, the force is roughly proportional to the amount of the bend. This force, directed away from the bat, is part of the force that propels the ball back toward the playing field. But when the bat bends so much that the stress on the wooden fibers on the side of the bat away from the ball (toward the catcher) exceeds the strength of the stretched fibers, the fibers break and no longer resist the incursion of the ball.* With less force directed toward returning the ball to the diamond, the velocity of the ball is much reduced, and instead of carrying over the infield, it sails feebly off the broken bat into an infielder's glove.

It is interesting to note that the heavier, thicker-handled bats of yesteryear were more rigid. Hence, they had higher natural frequencies and smaller amplitudes of vibration upon off-center impacts with the ball, stretched the wood fibers less, and hence broke less easily.

Since hickory has about twice the elastic modulus of ash, the stiffer hickory bats also vibrated with a smaller amplitude and higher frequency. (For a given configuration, the vibration frequency varies inversely with the square root of the density, and hickory is denser than ash. This reduction of the natural frequency from the greater density of hickory is not as important as the increase in frequency from its greater stiffness.) Since the energy stored—and then lost—in the vibration of the bat is proportional to the square of the amplitude, the stiffer, but heavier,

---

*The tensile forces are greatest when the bending is large—away from the node. But shear forces, which tend to separate layers of fibers, are greatest at the nodes where there is no bending. Both forces play a role in the breaking of the bat.

hickory bats will retain less of the collision energy for balls hit away from the sweet spot. Also important, the hickory bats and the thick-handled ash bats broke less often.

Although the increase in bat mass of a thicker-handled bat will slow the swing down somewhat, that effect is probably small. Hence, if a player is comfortable using a thicker-handled bat, he might gain a couple of hits on inside pitches over the course of a season. Now they won't break his bat, and he may add a few points to his batting average.

Indeed, in general, bats with thicker handles and bats with longer barrels have longer vibration-free zones of good hitting and are broken less easily by inside pitches. Edd Roush, hitting choked up with his 48-ounce, thick-handled hickory bat in the 1920s, probably never in his life stung his hands hitting an inside pitch and seldom, if ever, broke a bat. The long-barreled "bottle bats" used by players such as Heinie Groh* and Bucky Dent make sense—for some players.

ABERRANT BATS: ALUMINUM BATS

Hollow metal tubing bats—usually from aluminum alloys†—are now used in baseball played outside of the professional leagues. Bats constructed in strange and ingenious ways are used in base-

---

*Groh's bats had a barrel about 17 inches long, which then necked down abruptly to a rather thin handle that was also about 17 inches long; this odd bat was about half barrel and half handle. Groh swung the 46-ounce bat choked up with his hands spread slightly apart. He had small hands and preferred a thinner handle than was common in the 1920s. Dent used a more conventional bat, but with a long, full barrel that he swung choked up. Groh and Dent were largely line-drive hitters—though both hit important home runs. Indeed, Groh led the 1919 Cincinnati Reds—who beat the Chicago Black Sox in the World Series—in home runs (with 5) and in slugging average (at .431). Recently, some have used the term "bottle bat" differently—for any long-barreled, thin-handled bat.

†The metal bats that sustain the greatest distortion upon impact drive the ball farthest as the recovery of the distortion pushes the ball off of the bat. If the bat does not recover, it's dented and useless—and the ball receives no push and doesn't go anywhere. Considerable ingenuity is applied to finding materials and structures that are maximally flexible but still recover after large distortions—thus the use of special aluminum alloys and titanium. I will use the label "aluminum" for all metal bats.

ball variants, such as various forms of recreational softball, where there seem to be no special constraints on bats. The use of illegally altered bats has occasionally been detected in major league baseball. Questions arise concerning the efficacy of these aberrant bats.

The thin-handled bats made of light wood popular today are fragile, and not cheap. Hence, the cost of bats for amateur baseball is significant, given the economic scale of the sport on that level. Since bats made from aluminum tubing do not break, and can be made with appropriate balance and weight, these bats have become the bats of choice for such baseball and are now allowed by amateur rules. Today, it is hard to find a wooden bat among the rows of aluminum bats in most sporting goods stores.

The requirement that wooden bats be turned from one piece of solid wood places constraints on the size of light bats. Since woods less dense than ash are not strong enough to serve as material for bats, a light bat must perforce contain less wood and be smaller than a heavier bat. Moreover, the weight distribution along the bat is defined by the shape of the bat, and the traditional shape is that which makes the best use of wood.

Since the thickness of the aluminum tubing used to make the hollow aluminum bats can vary over wide limits, the weight and weight distributions of aluminum bats can be set almost independent of the size and shape of the bat. Currently, aluminum bats are shaped like the traditional wooden bats and are made to have similar weights and weight distributions,^{f} but this follows from convention rather than mechanical necessity. The freedom of design afforded by aluminum has, however, resulted in the manufacture of light bats with the full-size 2.75-inch barrels allowed by the rules of baseball, while wooden bats must have smaller barrels to keep to the same weight. Hence, the typical 32-ounce wooden bat will have a barrel diameter of only 2.50 inches. With the larger barrel, the 32-ounce aluminum bat will have about a 4 percent larger effective bat-ball hitting surface, with no counteracting disadvantage.

Aluminum bats may also be better than wooden bats in other

ways. The question has been raised as to whether a ball hit by an aluminum bat will go faster and farther than a ball hit by a wooden bat. Players also hold the view that an aluminum bat has a longer region of good hitting than a wooden bat. In particular, balls hit short on the bat—near the handle—are propelled more efficiently by an aluminum bat, and with less vibration and stinging of the hands, than by a wooden bat. And aluminum bats do not break. There appears to be some consensus among players that these reputed advantages of the aluminum bat are real. Are these conclusion correct? And if so, why?

I'll address first the question of the length balls are hit by an aluminum bat (or the speed with which ground balls hit by an aluminum bat travel through the infield). I have noted that the baseball is not very elastic. Even at the moderate velocities used in testing balls, upon striking a hard surface only about 30 percent of the collision energy is returned in the rebound. The return is less than 25 percent for a solid hit off a fastball. A baseball is more elastic than a beanbag, but not by much. The coefficient of restitution for a beanbag striking a hard surface is near zero; beanbags do not bounce at all. No matter how strong a batter is, he cannot hit a beanbag much past second base. Nevertheless, it is possible to design an implement that will hit a beanbag a long way. An examination of such implements can provide some insight into the mechanisms that allow baseballs to be hit farther by bats.

Though we will not be able to hit the beanbag very far with a baseball bat, we will do better with a tennis racket faced with a sheet of very thin rubber rather than with gut or nylon strings. With this contraption, in the collision of the beanbag with the racket, the beanbag will travel past the plane of the racket by perhaps a foot, stretching the rubber. Then the rubber will react, propelling the beanbag away from the racket with a rather high velocity in a kind of catapult effect. Actually, a sheet of rubber would induce too much air resistance for a really fast swing of the racket-bat, and we would do better by stringing the racket with very strong but quite elastic strings.

If we attach a long handle to the tennis racket and weight the racket head to copy the length and weight distribution of a baseball bat, and then swing the racket like a bat, we will be able to hit a beanbag the size and weight of a baseball quite a long way. In fact, could we find strings (perhaps made of steel springs) as elastic as gut or nylon and so strong that they would not snap under the large forces we would generate, we could hit this beanbag farther than a baseball struck by a bat! And indeed, we could hit a baseball farther with our racket-bat than with a regular wooden bat.

I explain this phenomenon by describing the (imaginary) observation of the results of a collision between a specific (also imaginary) racket-bat and the beanbag-ball (a baseball cover stuffed with beans so as to have the same weight as an official ball). The beanbag is flung toward the plate by a major league fastball pitcher and the racket-bat is swung by a major league power hitter. When the beanbag hits the racket face, the strings give way elastically and the bag moves 6 inches past the plane of the racket face. (This deflection is about ten times the compression of the baseball hit by a bat.) Then the strings react elastically, flinging the beanbag back toward the center-field bleachers.

The racket-bat-beanbag collision took place over a time of about $\frac{1}{100}$ of a second (rather than the time of $\frac{1}{1000}$ of a second in the collision of the wooden bat with the baseball) and the maximum force is 800 pounds (rather than the 8000-pound force which reversed the baseball flight). The 800-pound force compressed the soft beanbag-ball 1 inch—and distorted the racket-bat by 6 inches, as previously noted. Then about six-sevenths of the collision energy was stored by the racket and only about one-seventh by the beanbag. Since the beanbag is completely inelastic, all of the beanbag compression energy was lost to friction, but only 40 percent of the energy stored by the racket was lost. For the collision of beanbag and racket, more than 50 percent of the collision energy was returned to the beanbag as kinetic energy in the catapult-like action of the racket on the beanbag; the coefficient of restitution was greater than 0.700.

For a baseball, the energy return is only about 20 percent (corresponding to a COR of about 0.45 at this impact velocity.) A swing that would drive a baseball 400 feet using a wooden bat would drive the beanbag about 480 feet (and a baseball a few feet farther)—the racket-bat would hit a beanbag farther than the wooden bat would hit a baseball! And if the racket-bat were used on a baseball in Yankee Stadium, the ball could well be knocked clear out of the stadium—and not even Babe Ruth ever hit a ball out of that park.

Even an umpire as obtuse as fans claim in their calumny, would be suspicious of such a bat. But aluminum bats do have a little of the catapult-like elastic properties of the racket-bat, and surely *can* hit a baseball appreciably farther than can a wooden bat.

From measurements made by my colleague R. C. Larsen on the compressibility of aluminum bats, it seems that, for a given force, the distortion of an aluminum bat is about one-tenth[g] as great as the distortion of the ball (rather than one-fiftieth, as it is for wooden bats). Moreover, from the high frequency and the persistence of the sound emitted by the bat when stuck lightly at a node of the longitudinal vibration, it appeared that the distortion was quite elastic and that the compression energy would be restored quickly (unlike the longitudinal vibration.) Hence, in this mode, the bat was surely much more elastic than the ball, while the comparatively incompressible wooden bat stored little energy and was only slightly more elastic than the ball. The aluminum bat stored about one-eleventh of the collision energy in a highly elastic deformation of the bat that was returned efficiently to the flight of the ball, while the ball stored about ten-elevenths of the energy in the deformation of the ball, most of which was lost in friction.

Though the intrinsic elasticity of the aluminum bat will be near 100 percent, some of the bat-deformation energy will be retained in the energy of motion of the aluminum shell that makes up the bat. I estimate that 70 percent of the energy stored in the deflection of the aluminum may be returned to the ball. The elasticity of the ball at this impact velocity will be about 20 percent (the square of the coefficient of restitution at the relevant impact force). Using these values, 25 percent of the collision energy will be restored to the aluminum bat and ball, a value much greater than the 20 percent in the collision of the wooden bat and ball. This corresponds to an effective COR of 0.5 rather than the wooden-bat value of 0.45, and the home run hit 380 feet with the wooden bat will go more than 410 feet off the aluminum bat. The same kind of calculations suggest that a fungo hit 330 feet by a (regular) wooden bat would travel about 10 feet farther hit with the same swing by a similar aluminum bat.

Though the numerical values presented here can be considered only as a kind of illustrative estimate—in view of our uncertain understanding of the character of the aluminum bat—the aluminum bats now used for amateur baseball will surely hit a ball appreciably farther than will a wooden bat.

The second purported advantage of the aluminum bat is that it does much better than wood on inside pitches hit too near the handle. The aluminum bat is a kind of shaped hollow cylinder; since aluminum is about four times as dense as ash, to achieve the same weight the bat must be hollow. Aside from the greater strength of the aluminum, along the axis the hollow cylinder is more rigid than is a solid structure containing the same mass of material. The typical aluminum bat is about twice as stiff as a wooden bat of similar dimensions—about twice as much force is required to bend it a given amount. Hence, when a ball is struck badly—near an antinode at the handle or the end of the bat—the stiffer aluminum bat takes up less energy in longitudinal vibration than the wooden bat, and stings the hands less. Moreover, since the vibrational frequency is higher than for a wooden bat, the aluminum bat even returns some of that energy to the ball. Thus a ball mishit near the handle or the end of the bat will go farther off an aluminum bat than a wooden one; there is more room for error with an aluminum bat.

Perhaps more important, the aluminum bat does not break when a ball is hit near its handle. The force of resistance to the ball supplied in the course of the bending of the bat is not truncated by the breaking of the bat. With more force propelling the ball back, the ball will come off the aluminum bat faster—perhaps over the infielder's head—than off the broken wooden bat.

ABERRANT BATS: CORKED BATS

On occasion, players have modified their wooden bats by drilling a hole from the end of the bat along the axis and filling it with cork, cork balls, rubber, or other materials. A wooden cap is placed over the hole, then sanded and varnished to hide the modification. Neither the hole nor the addition of the filling are allowed under baseball rules.

It is desirable to determine whether such an illegal modification might create a bat that has properties that cannot be achieved by a legal bat—and hence might threaten the integrity of the game—or if the effect of the modification can only produce a bat with properties no different than might be achieved by a legal bat. It is not possible to conclude categorically that a specific physical change in a bat is good or bad. But it is possible to determine the character of the change and to determine if the change—if illegal—would produce a bat with characteristics outside of the range covered by bats constructed according to baseball rules.

Ordinarily, we describe a bat by its length and weight. With wooden bats all having similar shapes, that's a useful description. Two bats with the same weight and length will feel about the same when we swing them. But this simple description doesn't work so well in considering bats in which weights are added or subtracted illegally. Obviously, a couple of ounces added or subtracted at the end of the bat changes its character much more than taking away or adding the same weight in the handle would.

The properties of a bat relevant to swinging the bat so that it strikes the ball squarely are largely defined by the *moment of inertia* of the bat about the end of the bat and the pendulum length about the end. That moment of inertia is proportional to the resistance you feel when wagging the bat back and forth with your hands while keeping your hands in the same place. The pendulum length is the length of a simple pendulum of a bob weight on the end of a string that would oscillate with the same period as the bat suspended at the end. I call that length the "swinging length" of the bat, and define a "swinging weight" as the weight at that point that would give the same moment of inertia as the bat. Thus, if you had a long, almost weightless rod the length of the swinging length, mounted in a bat handle with a weight on the end of the rod equal to the swinging weight, this odd assembly would have the feel of a bat. Swinging it with your eyes closed, you would think that you were swinging a real bat with that swinging weight and swinging length.

For a typical bat 34 inches long weighing 32 ounces, I found a swinging length of 26 inches and a swinging weight of 28½ ounces. An interesting illegal bat modification is made by drilling an axial hole in the end of a bat and filling it with an extraneous material (e.g., cork, cork balls, rubber, rubber balls). The illegal modification was chosen to be a hole 1³⁄₁₆ inches in diameter drilled 6 inches deep along the axis from the barrel end of the bat. The hole was filled with cork. The density of the wood was 0.638—a typical density for American ash (or white ash)—and the density of the cork was 0.25 times the density of water. In this way, the weight of the bat was reduced by about 1½ ounces, centered about 31 inches from the handle end of the 34-inch bat.

This reduces the bat weight by an ounce and a half and, more important, the swinging weight by about the same amount. The swinging length was reduced by about three quarters of an inch. The bat has been changed significantly. It's lighter, and effectively shorter, and one can swing it more quickly. If a player is having trouble getting around on the fastball, this weight reduction will help him in very much the same way that a choice of a lighter bat might help. As with a lighter bat, the batter will probably not drive the fastball as far by 2 or 3 feet when he hits it well with the drilled-out bat, but he may hit it with good timing more often.

Generally, the player who drills out his bat stuffs the hole with cork or rubber. But this added material serves more as a detriment than an advantage. The extra material will add, perhaps, half an ounce to the weight of the bat—at the end—and then store about 2 percent of the energy of the bat. But that energy will not be effectively transferred to the ball. Even if the filler is quite elastic, such as superball rubber, its elastic energy cannot be transferred efficiently to the bat in the 0.6-millisecond bat-ball collision. Hence, the extra material will then only slow the bat down a little and slightly reduce the distance a ball can be hit. Such a filler will take another 3 feet off a 400-foot drive.

Whatever the advantages of the lighter bat, we can see that very near the same results achieved by the illegal modifications could be achieved legally by (a) choking up on the bat by about an inch, (b) using a bat with the same dimensions but constructed of slightly lighter wood, (c) sawing (legally) about three-quarters of an inch off of the end of the original bat, or (d) legally turning the barrel down from 2.50 inches to 2.40 inches. All of these modifications produce a bat that has almost exactly the same swinging weight and swinging length as the illegally drilled bat.

Solution (b) implies that a satisfactory "lighter" wood is available, which may not be the case. The American ash from which bats are made has an unusually high strength-to-weight ratio. Ash was celebrated in medieval times as the only proper wood from which to construct the lances of knights-errant; an ash lance was light enough to carry and wield and strong enough to impale the opposition. Any reduction in the weight of the wood seems to require a substantial compromise in strength and hardness.

The resonant frequency and position of the vibrational node are only important for mishit balls. And in any case, such illegal holes in the bat do not have a large effect on either the resonant frequency or the position of the node.

We have noted that aluminum bats are much more resilient than wooden bats and that in striking a ball they store energy in compression, which is then returned to the ball efficiently. It is conceivable that a device could be fashioned that could be inserted into a hole drilled in a bat and that could store energy upon impact, which would then be returned efficiently to the ball (in much the same way that the flexing of an aluminum bat stores, and returns, energy). Such a device could improve the effectiveness of a bat, but some reflection suggests that it would be difficult to design an effective device of this kind—and quite difficult, if not impossible, to make it hard to detect.

In summary: The characteristics of any specific baseball bat can be changed by drilling an axial hole in the end of the bat and filling the hole with some light, inactive, extraneous material. The modified bat differs from the original bat by its lighter weight and smaller moment of inertia. Bats drilled out in this way are excluded from play under the *Official Baseball Rules*. But the properties of such modified bats can be reproduced by a legal bat with the same "feel" and hitting characteristics.

TECHNICAL NOTES

a) In this model, we calculate the maximum velocity, v, of the bat from the relation

$$v = \sqrt{\frac{2E}{m + \epsilon^2 M}}$$

where m is the mass of the bat, M is the mass of the player, and E is a normalization constant approximately equal to the total energy produced by the muscles. The constant, ϵ, is taken as equal to ⅑, which is in accord with the observation that the maximum energy of bat and player are roughly equal. Using this relation, for a given total energy, E, we determine the variation of v with m for batters of a given weight, M. Then, as discussed earlier, we determine the velocity that a pitched ball will travel when struck by the bat of mass m moving with the velocity v.

In the approximation that the ratio of muscle mass to total mass is about the same for players of different size, we can expect that the energy put out by the muscles is proportional to the weight of the player. (For example, if the shape of a small player was the same as that of a player who was 30 percent heavier, the muscles of the larger player—10 percent longer and with a 20 percent larger cross section—would be expected to exert 20 percent greater force over a distance 10 percent greater to develop 30 percent more energy.) Hence, we can set E proportional to M and estimate how much farther a large player might be expected to hit a ball than a smaller player.

b) For a real, flexible bat, the signal of the impact travels along the bat as a transverse vibrational wave at a velocity that varies along the bat—very fast at the thick barrel, more slowly near the thin handle. Thus the signal from the impact of the bat will take some time to reach the hands. That signal velocity will be near the speed of the wave velocity of the highest important frequency wave. Taking this as the 3rd harmonic (the higher-vibration frequencies do not contribute much, as the collision damps them out), we estimate that velocity as $v_3 \approx \lambda_3 \times f_3 \approx 1750$ feet per second, where I take $\lambda_3 \approx 1.5$ feet as the wave length and $f_3 = 1180$ Hz as the measured frequency. Using that velocity, the signal will take about a millisecond to reach the hands from the point of bat-ball impact—and another millisecond for the effect of the hands to get back to the ball-impact region. With the duration of that impact along the order of half a millisecond, the handle and hands will neither receive the signal nor respond to the impact until the ball has left the bat.

This conclusion may break down for balls hit too near the handle, in which the signal has a shorter distance to travel and the ball-bat impact time is longer.

c) The position of the nodes of the fundamental vibration of a wooden bat can be easily determined with no more than the bat, a piece of string, and a light hammer. With the addition of a piano and a sense of pitch, one can also find the frequency. Suspend the bat by a string tied around its knob, tap the bat with a hammer, and listen. When the bat is struck at the end or where the barrel necks down to the handle, the tap will cause it to ring for a fraction of a second. As you move the position of the tap, you will find two points where there is no such ring—one near the trademark and one 6" or so from the handle end. These are the vibrational nodes. By matching the bat-ringing frequency with the notes struck on the piano, you should be able to determine the approximate frequency of the bat, which will probably be near 176 Hertz, or F in the octave below middle C.

This procedure does not work well with an aluminum bat, as the striking of the bat anywhere incites a vibration of the column of air in the hollow bat. This organ pipe–like vibration will be near the same frequency as that of the fundamental bat vibration, and will obscure it.

There are higher vibrational frequencies with five and more nodes, but these are not excited efficiently by the ball impulse, which is longer than the period of these high-frequency vibrations.

When the bat (wood or aluminum) hanging from two feet of string tied about its handle is struck a little harder well above the trademark, the handle end of the bat will jump in the direction of the hammer blow, as well as vibrate. An impact of the ball at the same place on the bat held by the batter then drives the bat into the back hand of the batter (the right hand for a right-handed batter), which will sting from the 170-Hz vibration. But if the bat is struck on the end of the bat, the handle will jump in the opposite direction, to sting the leading hand. There will be a point in between where the handle does not react at all when the bat is struck. This is the *center of percussion*. For a baseball bat, the center of percussion is usually near the vibrational node—hence, the no-sting point—and in the vicinity of the trademark.

d) The basis of the identification of the sweet spot with the node of the fundamental vibration can be found in the neurophysiology of sensation. Different modalities of somatic pressure sensation are detected by different receptors, which send signals through different dedicated nerve systems to different sites in the brain. In the hairless skin of primates (such as that on the batter's hands), vibrations near frequencies of 200 Hz are sensed by sub-cutaneous Pacinian corpuscles. These are not sensitive to lower-frequency pressures and are not very sensitive for vibration frequencies above 300 Hz. Thus, the higher-frequency bat oscil-lations, though they do take up energy and reduce bat perfor-mance when they are excited, are not detected by the corpuscles. Only the fundamental vibration stings the hands. Thus the sweet spot, as defined by baseball batters, is determined by the absence of the unique sensation derived from such signals, and that absence occurs when the ball is hit at the node of the fundamen-tal bat vibration—and nowhere else. This node is near the point at which the ball comes off the bat fastest and travels the farthest and is thus identified in the batter's experience with good hitting.

Just as the node of the fundamental vibration can be deter-mined by a simple home experiment, the position of the sweet spot can be found through an equally simple experiment that can be performed by anyone who has ever played baseball. Hold a wooden bat as if you were batting in a game, and strike a rela-tively unyielding narrow vertical structure firmly with the barrel of the bat at various points in the hitting zone. A doorjamb will work, but the 4-inch-diameter concrete-filled steel "Lally columns" found in the basements of many American homes are even better. At one impact point it will feel *just right*, and you will have found your sweet spot. It will be the same point as the fundamental node that you measured.

e) Although most calculations of physical quantities require some level of approximation—the mathematical model is almost never a precise map of reality—Professor Van Zandt's calcula-tions of the characteristic vibrational frequencies and amplitude shapes for a baseball bat should be accurate. However, his calcu-lated frequencies were about 25 percent lower than those meas-ured by Professor Uwe Hansen. With that correction (which presumably followed from a difference in the elastic properties of wood used for bats), the calculated frequency of the bat Van

Zandt considered was 170 Hz and the first harmonic was 560 Hz.

f) In fact, manufacturers seem to make aluminum bats with somewhat different weight distributions than wooden bats. Since the weight of a section of the hollow bat is in the shell, or skin, of the bat while the mass of a similar section of a wooden bat is distributed through the whole section, there is a tendency to place relatively less weight in the barrel of the aluminum bat and more in the handle. If the same gauge (or thickness) of aluminum were used throughout the bat, a 1-inch section of a 2.75-inch-diameter barrel would weigh about 2.75 times as much as a 1-inch section of the handle. For a wooden bat of the same dimensions, the ratio would be much greater—about $(2.75/1.0)^2$ $= 7.56$. Usually, the manufacturers only partially counter this by using thicker material in the handle than in the barrel of the aluminum bat.

The freedom of design of aluminum bats has the consequence that nominally similar bats may have quite different properties— that is, different swinging weights and different swinging lengths. Hence, the comments I make about a specific aluminum bat may not apply exactly to all aluminum bats.

The different weight distribution of the typical aluminum bat has both advantages and disadvantages. With less weight in the barrel behind the impact with the ball, the 32-ounce aluminum bat acts in the collision with the ball more like a 30-ounce wooden bat, transferring less energy to the ball and costing some distance loss on a solid hit. But the trampoline action of the bat acts to drive the ball much farther than a wooden bat can anyway, and the effectively lighter bat can be swung a little more quickly. Also, since the extra weight in the handle increases the turning moment of the bat, a ball mishit too near the end of the bat or the handle might lose a little less energy in the turning of the bat if the bat was quite stiff and loosely held by the batter. But this advantage is, at best, small, and the reduction of vibration by the stronger handle of the aluminum bat is a more important factor in increasing the length of the good hitting region.

g) The measurements of the compressibility of the balls and bats were made statically and can only approximate the actual forces generated by the collision of the spherical ball with the cylindrical bat. We believe the results are more likely to underestimate than overestimate the distortions of the bats.

RUNNING, FIELDING, AND THROWING

RUNNING

Baseball players run. They run the bases, and they run in the infield and outfield in fielding balls. To understand this split-second sport, we must have some understanding of how *fast* they run.

We can begin to understand how fast ballplayers run by reviewing how fast world-class sprinters run on nearly ideal tracks—their speed is surely a limit for baseball players. Figure 7.1 shows record times for sprints in 1998, where the straight line fits a simple model that can be simply described by noting that a world-class track athlete like Carl Lewis, who won the 100-meter dash in the 1984 and 1988 Olympics, running with track shoes on prepared tracks, can sprint at a rate of about 38 feet per second after a start, out of starting blocks, that takes 1.235 seconds.* An equally fast man, given a running start from 30 yards behind Lewis so that he crossed the starting line as the starter's gun went

---

*The extrapolation of world-record times given by the straight line in the graph predicts a time for Lewis, on a track, of 4.4 seconds for a 40-yard dash. Hence, many of the 40-yard-dash times reported from football training camps must be taken with considerable reserve.

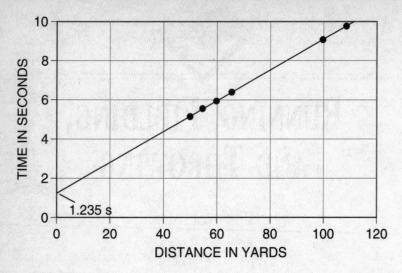

FIGURE 7.1: *The points show world-record times for sprint distances in 1998. The straight line is a best linear fit to the points.*

off, would be about 1.235 seconds, or 15 yards, ahead of Lewis at the 30-yard mark.

I take the speed of a very fast outfielder, running on grass in a baseball uniform with baseball shoes and carrying a glove, as 30 feet per second. But how much time does he lose in starting? Unlike Lewis, the outfielder generally doesn't even know in which direction to run at the moment the bat hits the ball. He can probably tell whether to go left or right after the ball goes about 40 feet, which takes about 0.3 seconds. If he is stationed 300 feet from home plate, he will first hear the crack of the bat about that time. But not only doesn't he have starting blocks, he is not even leaning in the direction he has to go—he can't even know what direction that is. I would have to conclude that even Cool Papa Bell, perhaps the fastest man to ever play the outfield and who could probably run almost as fast as Carl Lewis, would lose 1.75 seconds on his start in the outfield.

On the bases, without gloves and free to swing their arms, the speedier players will run a little faster—say 33 feet per second—and will be able to start a little more quickly. But, if the base runner is not to be caught leaning by the pitcher's pickoff throw, even the quickest will still not start as fast as Lewis. I estimate that Lou Brock, who stole 938 bases in his long career, still lost 1.5 seconds upon starting.

With these recipes, we can calculate the distance the fastest players can cover in a given time; those distances vs. time are shown in Figure 7.2, along with distances Carl Lewis might attain. The exact values are not too reliable for short times, but are probably good estimates for times over 2 seconds.

JUDGING FLY BALLS

How do players judge fly balls? Just what did Tris Speaker, Richie Ashburn, Mays, and Mantle, as well as the kids of my youth (and Little Leaguers today), do to "judge" a fly ball? How do players know where to run to catch a ball that is sailing toward them? An

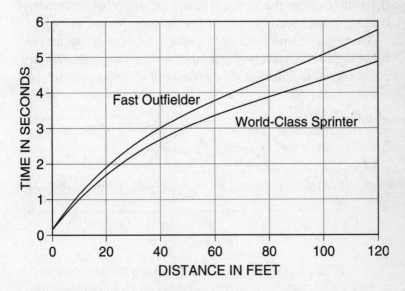

FIGURE 7.2: *The upper curve shows my estimate of the distance vs. time for a very fast baseball player in the outfield, while the lower curve shows my prediction for a world-class sprinter in a race on a prepared track.*

android with a computer brain might calculate trajectories, but I suspect that humans follow simple learned patterns derived from experience; a human can beat supercomputers on pattern recognition. But to compute, or to operate from a set of memories, the player, human or android, must work from information provided by the senses. Hence, without knowing just how the information the fielder receives is processed by the brain, we can still discuss the limitations on that information and understand much about judging fly balls.

To discuss, quantitatively, running down fly balls, one should have some idea as to how long balls stay in the air and how much ground players can cover in that time. The trajectories of the balls and their flight times can be calculated rather accurately from the known air resistance of the balls; some fly ball trajectories are shown in Figure 2.3. And with the aid of Figure 7.2, we can estimate how far outfielders can run during those flight times. But how does the player decide where to go? How does he determine where the ball will come down? What information does the player need to estimate the distance the ball will go?

We begin our answer to that question by considering the ball hit straight at the player—the most difficult hit to judge. Our first conclusion is that the initial rate of takeoff of the ball—as seen by

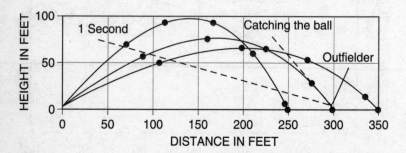

FIGURE 7.3: *The trajectories of three fly balls hit different distances directly at an outfielder. Markers show the position of the ball after 1, 2, 3, and 4 seconds. The line of vision of the outfielder at the 300-foot mark is shown by the broken line. A second broken line shows his line of vision as he prepares to catch the descending ball.*

the fielder—doesn't help; balls hit different distances can look the same initially. Figure 7.3 shows the trajectory of three fly balls where the position of the ball is shown by markers every second. One represents a fly ball that will be caught by the motionless outfielder stationed 300 feet from home plate. The other trajectories represent balls that drop at the 250-foot mark and travel out 350 feet from home plate, both to be caught, hopefully, by the outfielder after a run. All of the balls stay in the air between 4.3 and 5 seconds. So the fielder must decide to run in, run out, or stay put. But from his position, the changes in the line of sight of the three balls as they leave the bat are almost identical for the first second, during which the angle of rise seen by the fielder is about 15 degrees for each of the three fly balls.*

Hence, the player can't tell from the initial rate of rise alone whether he should run in or out. The early differences in the trajectories are insufficient clues. Later, the trajectories differ substantially. After two seconds, or a little less for a very perceptive player, the outfielder knows whether to run in or out from his assessment of the line of flight—which now is appreciably different for the different balls. But, having hesitated about two seconds, he can't cover much ground; he won't get a good jump. If he waits a couple of seconds to start running, it is clear from Figure 7.2, he won't cover the 50 feet he needs to in order to catch the short fly that lands after another 3 seconds. And the long ball that lands after another 2.3 seconds will be bounding toward the fence by the time he reaches its landing point.

Of course, the experienced outfielder uses other clues. He watches the swing—is it a wheelhouse power swing or an off-balance chop by a batter half-fooled by the pitcher? How did the hit sound? Was it the "crack" of the squarely hit ball or the "clunk" of the mishit? Perhaps he unconsciously estimates the path of the ball from its perceived size—at these distances, depth perception follows from perceived sizes and not from binocular effects. All of

---

*The three trajectories that are described here were taken as examples from a large set of fly balls that had the same rise in the first second from the outfielder's perspective.

these data are used by the player to construct a pattern that, from experience, he can use to estimate how far the ball will go, and hence whether to run forward or backward. But these data can be hard to read, and a mistake can be very expensive.

Let us consider the ball in Figure 7.3 that is hit over the fielder's head so as to land 50 feet behind him 4.3 seconds after leaving the bat. If the player guesses almost immediately that this is a long ball and turns and starts running back, he will reach the ball after an easy lope. But if he misjudges the ball, thinking it will fall short, and takes but one initial step forward, he is in trouble. For he will have to move forward and then stop before starting back, and the stop-and-restart time will cost him about 1.8 seconds. Moreover, he will have to retrace the 5 feet he moved forward, and will thus have to cover 55 feet, not the original 500. We see from the graph of Figure 7.2 that he won't come close to the ball in the 2.5 seconds that are left, and a ball he should have caught will sail over his head for a double.

The opposite misjudgment, of the short, high fly—in my grandfather's day, a "can of corn"—hit so as to land in front of the outfielder about 250 feet from home plate after a flight of 5 seconds through the air, carries a similar penalty. If the ball is judged correctly, it should be caught easily after a 50-foot jog. But, again, a too-precipitous misjudgment can be disastrous. If the player starts back and then must turn and run forward, he won't quite get to the ball and it will fall for a single.

For the well-played fly, during the time the ball is in the air the fielder's estimate of the distance it will carry is continually being refined so that he will be near the correct spot when the ball is about a second from the ground. At that point, alignment procedures take over, directing player and glove to a precise conjunction with the ball. I doubt that the brain does much in the way of calculation, but I expect that it compares the new pattern with its storehouse of patterns from the experience of past catches. Hence, the value of experience; hence, the need for practice.

During the last alignment second, the player tries to position himself so that the falling ball moves exactly at him, as shown in

Figure 7.3. If the line of sight of the ball drifts upward, the ball is passing over his head and the player moves back; if the line of sight falls, the ball will land in front of him and he trots forward.

Balls hit to the left or right are much easier to judge, and errors in estimated distance are punished less severely. With the ball hit to the side, the fielder can see that it is to the left or right in a few tenths of a second. Then, while he begins to run to the left or right, he can observe the trajectory of the ball much as the reader looks at the pictured trajectory of Figure 7.3—and can judge the trajectory just as the reader does.

With the left-right question solved in the first half-second of flight, how does the fielder proceed? Whether it is teenagers shagging softball flies on a playground or major leaguers fielding fungoes in spring training, we see that almost all balls are caught on the run. For a high, lazy fly hit to land not too far from the player's position, the run is an easy lope timed so fielder and ball meet as the ball returns to about the height of the player's eyes, five or six feet above the field. If the ball is a line drive with little hang time, or a long drive to the warning path, the player may have to run very hard, but, again, he and the ball meet at the point of the catch. But how does the player plot the course that results in that precision?

Part of the answer is that the player runs laterally so as to keep the left-right motion of the ball aimed in his direction. Of course, he knows how to catch balls thrown at him. Though the head of the player moves erratically as he runs, his sensory system takes that motion into account and he can still determine whether the ball is moving at him or drifting backward or forward.[a] If the ball is hit initially to the fielder's right, and in the course of the running catch the ball drifts to the right, the fielder will speed up; if the ball drifts to the left, the fielder will slow down. The direction or angle of the run is more difficult to set but is designed such that the falling ball—somewhat after it reaches the apex of its flight—moves, again, almost exactly at the player. Later, when the ball is coming down, if the ball drifts up too fast, the running fielder will turn toward the fence; if the ball drifts downward, the fielder will turn toward the infield. Hence, for nearly a second before the catch, the ball is coming right at the moving fielder, with no motion to the left or right and hardly any motion up or down. Therefore only one pattern need be learned, a pattern that is valid no matter how fast the player is running—if he is running at all.

We can gain some insight into the process by reviewing Figure 7.4, which shows a plausible scenario for the running catch of a long fly ball. We take the ball trajectory as that of the long fly shown in Figure 7.3. Since the fielder does not know how far the ball will go for at least a second and a half, he will start directly to his left. He will be able to determine that the ball will go left in about 0.3 seconds. Then, after 1.5 or 2 seconds, he will be able to judge the distance and will begin to run deeper, following the path shown as the solid line in Figure 7.4 and moving at a speed such that he always sees the ball moving directly at him as far as left-right motion is concerned. Hence, ideally, at each second interval, he is looking at the ball in the direction shown by the dotted line of Figure 7.4; in the system of the fielder the ball's left-right movement is null. Of course, at the beginning of his run he will be a little behind the ideal position, but he can detect that lag and speed up.

To catch this long ball, hit about 350 feet at an angle of 10 degrees with respect to the line from the fielder to home plate in the direction of the right-field power alley, the right fielder must run about 70 feet in about 4.3 seconds; from Figure 7.2, we can see that this is within the range of the average outfielder. Note that the fielder may even slow down at the end of his run, as the ball is now traveling slower.

But even for balls hit to the left or right, it's not easy to know how far the ball will go, especially in the first second of flight, and the fielder may make a mistake. Let us assume that he misjudges the ball seriously and starts in slightly toward home plate, only later correcting himself. Now he will have to run an extra 10 feet or so, though a fast outfielder will probably still be able to make the catch.

In summary, the misjudgment that changes ordinary running

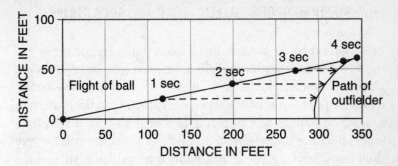

FIGURE 7.4: *The circles show the left-right position, at 1-second intervals, of a long fly ball (as shown in Figure 7.3) hit 350 feet toward the power alley. The solid line shows the position of a right fielder making a running catch where the player, whose initial position was 300 feet from home plate, was not sure initially how far the ball would go. The lateral dotted lines show the line of sight of the player watching the ball at 1-second intervals.*

catches into more difficult but still makeable catches for a ball hit 10 degrees away from the fielder will change easy catches into base hits for balls hit directly at the fielder. Stopping and turning around is much more expensive in time than running to the side at the wrong angle.

Of course, not all running catches are made as just described. The experienced outfielder may quickly judge where the ball will land, and run there and wait for it—but in fact, if the ball is well within his range and there are no base runners, he will usually catch the ball on the move, as per the recipe I have described. If the ball looks as if it will travel to the limit of his range, he may put his head down and run as fast as he can to where he expects the ball to land, and only look up at the last moment. The old-timers prided themselves on their ability to turn and run to the long ball without looking back, but you seldom see that today. And if there are men on base and the ball is easily in his range, the fielder may take a path that allows him to catch the ball running in so that he can quickly get off a powerful throw to the infield. So as we see, there are common variations, but for the most part players catch the ball on the run, just as they did as boys when they first began to play the game.

FIELDING GROUND BALLS—LONG AND SHORT HOPS

Artificial turf covers the infields of many major league parks, giving true bounces to ground balls. The grass infields of the other parks are tended carefully, so there too there are few erratic bounces. But players who learned the game in sandlots have had to learn to handle bad bounces. The fielder's rule is simple to state, albeit not always simple to execute: Field the ball on very short hops or long hops; avoid the intermediate. An average ground ball moves, perhaps, at 60 mph, or 88 feet per second. With a reaction time of 150 milliseconds—plus 50 milliseconds more to move the glove—a shortstop can correct for the bad hop of a ball that hits 20 feet away or more, but he will be handcuffed on shorter bad hops. Of course, if the ball hops only a few feet, even if it is deflected a little, it will still hit the glove close enough to the center to lodge securely. Hence, if the infield is imperfect, short or long hops are secure, but players just don't have time to adjust to bad hops of balls that hit 5 to 20 feet in front of them, and those bad hops are destined to be marked as errors by the scorekeeper.

THROWING IN THE INFIELD AND OUTFIELD

We know that some pitchers throw harder than others; the variance in the velocity of the fastballs of different major league pitchers is more than 10 percent. A few of the fastest pitchers, such as Walter Johnson, Bob Feller, Lefty Grove, Slim Jones, Nolan Ryan, Roger Clemens, and Randy Johnson, threw or continue to throw balls crossing the plate as fast as 95 mph. Conversely, a few very good pitchers, such as Vic Raschi with the Yankees in the '30s and '40s and John Tudor in the '80s, threw fastballs that crossed the plate traveling no more than 85 mph.

Without the benefit of measurements, we can assume that position players throw with similar velocities. Since a good "gun" is an absolute necessity for catchers, third basemen, and short-

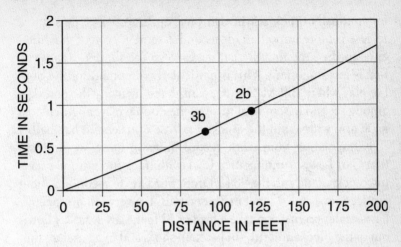

FIGURE 7.5: *Elapsed time vs. distance for a ball thrown with an initial velocity of 100 mph. The distances from home plate to third base and from home to second are marked.*

stops, we need only consider differences at these positions between good and very good arms. Figure 7.5 shows the elapsed time as a function of distance for balls thrown with an initial velocity of 100 mph—a throw from a player with a very good arm. Let us assume that the difference between the "guns" and merely good arms is about 5 percent, where the fastest deliver the ball with an initial muzzle velocity of about 100 mph. With such a delivery, a ball thrown 135 feet by the shortstop or third base-man to the first baseman's glove will take about 1.09 seconds. A player who throws the ball with an initial velocity of only 95 mph will deliver the ball to the first baseman about 0.06 seconds later. During this time the fastest runners will have covered about 2 feet, and we know that 2 feet is significant.

Of course, if a player with only a moderately good arm can release the ball 0.06 seconds more quickly, he can make up for the advantage enjoyed by the "gunner." He can't gain much on the actual throwing—which takes about 0.10 seconds for all play-ers—but players generally take about 1 second from the time the ball sticks in their glove to the beginning of their throw, and an exceptionally quick player can pick up time there. This quick release is more important for second basemen, even as the difference in the time that the ball takes to go the shorter distance to first is less important. This is particularly important for the double play where Bill Mazeroski, considered by many the best ever among second basemen at turning the double play, augmented a good arm with about the quickest release ever seen in baseball.

Extrapolating from world-record times in track, we estimate that Carl Lewis, running on a track with track shoes from a starting block, could run the 90 feet from first base to second in about 3.6 seconds—and 78 feet in 3.3 seconds. If a very fast prospective base-stealer on first can take a lead of 12 feet, start as fast as Lewis out of the blocks and run (in baseball shoes and in a baseball uniform) just 10 percent slower, he will slide into second 3.5 seconds after a dead start—if he loses nothing on the slide. Every 3 feet of lead is worth about a tenth of a second, and a rolling start is worth a good half-second. Indeed, the difference between the runner's having his weight on his front foot rather than his back foot (but don't let the pitcher catch you leaning!) must be worth more than a tenth of a second; hence, the cat-and-mouse play between the pitcher and the runner on first.

A pitcher's fastball takes about 0.45 seconds to arrive in the catcher's mitt; a slider will take perhaps 0.03 seconds longer, allowing an extra foot for the base runner; and a curve will take an additional 0.13 seconds, translating to 4 feet for the runner. But the throwing motion—defined as the time after which the pitcher, throwing from the stretch, can't change his mind if he wants to escape a balk—is about 0.8 seconds to the release of the ball. Hence, with runners on the total time from stretch to catcher (no windup with men on base!) for a fastball is about 1.3 seconds, through some pitchers are as much as 0.2 seconds (or 6 running feet) slower. And a few relief pitchers seem to be significantly faster—perhaps by as much as 5 running feet.

If the catcher throws with an initial velocity of 100 mph (which is about the absolute limit), the ball will arrive at second base, 128 feet away, about 1.0 seconds after his release. But like the pitcher, he needs time, typically about a second, to get the ball out of his mitt and complete the throwing motion. Adding the times together—1.3 seconds for the pitcher to throw a fastball to the catcher and 2.1 seconds for the catcher to throw to second base low and on the first-base side of the bag in place for the tag—we have 3.4 seconds for a perfect play and the base runner is out. But give a fast man a good start—worth about 0.2 seconds—and no one can throw him out. If the throw is only slightly off, the runner has an even better shot at stealing; for every foot the shortstop or second baseman has to move to make the tag, the runner gains another 2 feet or more.

Some say it is easier to steal third than second, but since the tactical gain is not as great, stealing third is relatively uncommon. The throw to third is shorter and the ball gets there about 0.7 seconds after the catcher's release, or about 0.3 seconds sooner than the throw to second. And that means about a 10-foot difference for the runner, who can also take a larger lead at second. If he can take a lead longer by more than 10 feet than he could take at first, stealing third is easier than stealing second.

Since outfielders' throws cover greater distances, the differences in the throwing velocities of outfielders leads to considerable differences in balls' times of flight, with significant tactical effects on the game. Moreover, hitting plays a somewhat larger role in the whole craft of the outfielder than it does for second basemen, shortstops, and catchers; hence, outfielders with relatively weak arms—like Ralph Kiner—are tolerated if they can hit (Kiner, who hit 369 home runs in 10 seasons, was far from merely "tolerated.")

Because outfielders throw on the run, they can probably throw with muzzle velocities that run as high as 10 mph greater than those of infielders handling normal chances and throwing from set positions. Since throwing plays a less central role than some other skills, there is probably also a greater variance in the throwing velocities of outfielders than there is even for pitchers. I would expect that Roberto Clemente, throwing on the run from right field, let the ball loose traveling about 110 mph, whereas some of Kiner's throws from left field probably started out with velocities no greater than 90 mph. (Here let me note that in a domed stadium with no wind, the player who throws the ball *only* 90 mph can send the ball about 310 feet in the air, while the rare player who can reach 110 mph can throw the ball from home plate to bounce off of the center-field wall 400 feet away.)

Although there is a variety of tactically important throws where velocity is important, I choose to examine just one special, and especially dramatic, throw: the one designed to catch at home plate a runner who has tagged up after a catch. In my scenario, it is the last of the ninth and the score is tied, so the outfielder does not have to hit a cutoff man. I have the fielder planning to catch the ball on the run and his throw starting at a point 300 feet from home plate. Moreover, I assume that he plans—wisely or not—to hit the catcher on the fly, with the catcher's mitt on the ground, ready to make the tag.

Figure 7.6 shows the different trajectories that must be used by players throwing the ball with different initial velocities. The man who throws very, very hard, with a velocity of 110 mph, throws "on a line" with a low trajectory (and at an elevation angle of about 13 degrees); the man who has a relatively weak arm and who lets go of the ball with a velocity of 90 mph must throw the ball with a substantial arc (and an initial angle of elevation of about 27 degrees). He loses time not only because his ball is mov-

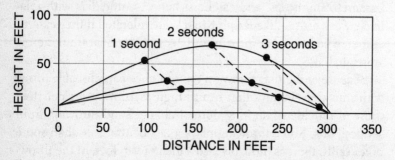

FIGURE 7.6: *Trajectories of throws with different initial velocities that travel 300 feet. The solid markers show the position of the ball at 1-second intervals.*

ing slower but because it traverses a high arc that is nearly 10 percent longer than that of the stronger-armed outfielder.

Very few players can go from third to home after a catch in less than 4 seconds; and of course, a runner can't start until the ball is caught. The differences in time of flight between hard and weak throws is thus critical. From Figure 7.7, we see that the very strong throw gets to the plate in about 2.65 seconds. Thus, if the player takes 0.6 seconds after the catch to release the ball, his throw will reach the catcher about 3.25 seconds after the catch and even the fastest runner will be out. However, the man with the merely good arm, throwing with an initial velocity of 100 mph, will put the ball in flight for about 3.1 seconds, and the ball will get to the plate about 3.7 seconds after the catch. If the throw is perfect, the catcher will probably tag out even the fastest base runner, but if the throw is even a few feet off, the catcher will be hard put to tag the runner before he touches the plate. And the ball from the 90-mph thrower will get to the plate in no less than 4.4 seconds after the catch, and the game will be lost. Of course, in all of this I assume that the ball to be fielded is a soft, high fly, so that the fielder can get a bead on it and catch it while running in toward the plate. If the catch is made while the fielder is stationary, one would have to add as much as 0.5 seconds to account for a slower release and a slower ball.

So 300 feet is too far away in my scenario. And, indeed, most teams station their outfielders no farther than 250 feet away under these circumstances. After all, if the ball is hit much farther than 250 feet, the game is lost anyway.

Let us note that those with weaker arms might be better off throwing at a lower angle and getting the ball to the plate on a bounce. If the surface is artificial turf, the 90-mph player can gain as much as 0.2 seconds, or 6 feet, on the runner by throwing on the bounce. But if his team is playing on grass and his groundskeepers have kept the grass long and well watered to help their team, which relies on singles, speed, and baserunning, the ball may lose so much speed on the bounce that nothing will be gained.

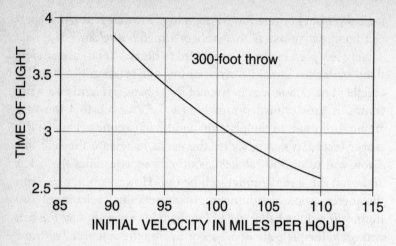

FIGURE 7.7: *Flight times for balls thrown 300 feet with different initial velocities.*

TECHNICAL NOTE

a) Just as a higher animal, predator or prey, must be able to tell if another animal is moving directly on an interception course with him when they are both running, humans can make that differentiation quite precisely. If the head is stationary, the eye will orient itself in the head so that the image of a moving object, such as a ball, will be centered on the fovea, the optical center of the retina. If, in time, the image of a ball stays so centered on the motionless eye, the trajectory must be such that the ball will intercept the player (and if he waits too long, hit him in the eye!). If the ball is not coming directly at the fielder, he must move his eyes to keep the image centered at the fovea. That necessary eye motion tells the player that the ball is not coming at him and that he will have to move if he is to catch it.

Remarkably, this result holds when the player is running with his head bobbing and moving erratically. With this motion, the eye watching the ball would have to move in the head to track the ball even if the ball was stationary. How can the player correct for the motion of his head so that the motion of the ball image caused by the ball's flight can be determined? The player, aided by hair cells in the vestibular canals of his ear that act as accelerometers, accurately records the motion of his head via his brain-computer, which integrates the accelerations to determine the head-motion velocities. The brain than subtracts these velocities from the total eye motion, leaving those motions of the eye that follow from the character of the player-ball trajectory to tell the player what he should do to catch the ball.

I have found that a simple experiment illuminates this remarkable motion correction. Move your head back and forth about once a second with an amplitude of about 20 degrees while you are reading this book. You will probably still be able to read, though with a little difficulty; your eyes continue to point at the words on the page as your sensory system corrects for the head's motion. Then hold your head steady and move the book back and forth with a matching amplitude and frequency. Now reading is much more difficult.

INDEX